The Dadant System of Beekeeping

by Charles P. Dadant

with an introduction by Jackson Chambers

This work contains material that was originally published in 1920.

This publication is within the Public Domain.

This edition is reprinted for educational purposes
and in accordance with all applicable Federal Laws.

Introduction Copyright 2018 by Jackson Chambers

COVER CREDITS

Front Cover
Honey Bee Grooming by Oasalehm (Own work)
[CC BY-SA 4.0 - https://creativecommons.org/licenses/by-sa/4.0],
via Wikimedia Commons

Back Cover
Honey comb by Merdal at Turkish Wikipedia
[GFDL - http://www.gnu.org/copyleft/fdl.html],
or
[CC-BY-SA-3.0 - http://creativecommons.org/licenses/by-sa/3.0]
or
[CC BY-SA 2.5-2.0-1.0 - https://creativecommons.org/licenses/by-sa/2.5-2.0-1.0],
via Wikimedia Commons

Research / Resources
Charles P. Dadant Information
Dadant & Sons, Inc.
https://www.dadant.com/history/

Wikimedia Commons
www.Commons.Wikimedia.org

Many thanks to all the incredible photographers, artists,
researchers, and archivists who share their great work.

PLEASE NOTE :
As with all reprinted books of this age that are intended to perfectly reproduce the original edition, considerable pains and effort had to be undertaken to correct fading and sometimes outright damage to existing proofs of this title. At times, this task can be quite monumental, requiring an almost total rebuilding of some pages from digital proofs of multiple copies. Despite this, imperfections still sometimes exist in the final proof and may detract slightly from the visual appearance of the text.

DISCLAIMER :
Due to the age of this book, some methods or practices may have been deemed unsafe or unacceptable in the interim years. In utilizing the information herein, you do so at your own risk. We republish antiquarian books without judgment or revisionism, solely for their historical and cultural importance, and for educational purposes.

Self Reliance Books

Get more historic titles on animal and stock breeding, gardening and old fashioned skills by visiting us at:

http://selfreliancebooks.blogspot.com/

introduction

Here at **Self-Reliance Books** we are dedicated to bringing you the best in *dusty-old-book-knowledge* to help you in your quest for self-sufficiency and food independence.

This special edition of **The Dadant System of Beekeeping** was written by Charles P. Dadant, and first published in 1920, making it just shy of one-hundred years old.

Dadant immigrated to the US from his native France when he was in his mid-forties, and settled in Illinois. He was once the largest producer of honey in America. The Dadant Family have been in the Beekeeping business now for over 150 years.

The book features chapters on *Early Experiments, Natural History, Size of Hives, Drones and Drone Production, Apiary Management, Handling Bees, Nomadic Beekeeping, Wintering, Diseases of Bees,* plus more.

A fabulous old book written by the namesake of the system. An absolute *must*-read for all those starting out in, or even considering taking up Beekeeping.

~ *Roger Chambers*
State of Jefferson, March 2018

CHARLES DADANT

CONTENTS

		PAGE
CHAPTER 1.	Early experiments—Natural History.	1
	The Queen	5
	The worker-bee	6
CHAPTER 2.	Size of Hives.	9
	The large hive	10
	Small hives	14
	Safety in wintering	20
	Frame spacing	21
	The supers	23
	Side storage	26
	Queen excluders	26
CHAPTER 3.	Drones and Drone Production.	29
CHAPTER 4.	The Dadant Hive	35
	A simplified Dadant hive	42
	Hive making	46
CHAPTER 5.	Handling Bees.	47
CHAPTER 6.	Our Apiaries.	49
	Outapiaries	52
CHAPTER 7.	Apiary Management.	55
	Spring	56
	The honey crop	57
	Increase	57
	Queens for increase	58
CHAPTER 8.	Swarm Prevention and Supering.	65
	Prevention of natural swarming	65
	Putting on supers	72
CHAPTER 9.	Extracting.	75
	Extracting implements	80
	Robbing	85
	Varying honey crops	87
	Requeening	88
	Queen Introduction	89
CHAPTER 10.	Nomadic Beekeeping.	91
CHAPTER 11.	Fall Management.	97
CHAPTER 12.	Wintering.	99
	Cellar wintering	100
	Our house cellar	101
	Wintering in clamps	102
	Wintering out-of-doors	102
CHAPTER 13.	Diseases of Bees.	107
CHAPTER 14.	Enemies of Bees.	113

ILLUSTRATIONS

Charles Dadant	Frontispiece
Moses Quinby	Fig. 1
L. L. Langstroth, inventor of the movable frame hive	Fig. 2
C. P. Dadant, sons and son-in-law	Fig. 3
Head of Queen	Fig. 4
Head of worker	Fig. 5
Frames of Langstroth and Dadant-Quinby size	Fig. 6
Twenty-frame Dadant hive tried by Charles Dadant	Fig. 7
Coffin-shaped frames tried by the senior Dadant	Fig. 8
Eight-frame Langstroth and Dadant hive side by side	Fig. 9
Several stories of eight-frame hives make too tall a pile	Fig. 10
Shallow brood-chambers, same depth as supers	Fig. 11
Ekes or nadirs of by-gone days	Fig. 12
The DeLayens long idea hive	Fig. 13
A regularly laid and well filled comb of worker brood	Fig. 14
Comparison of Langstroth and Dadant Supers	Fig. 15
Queen excluders exclude ventilation and free access to supers	Fig. 16
Modern excluders are better than the old, but are still in the way of the bees	Fig. 17
Comb built without foundation or with only a small starter	Fig. 18
The drone trap should not be used except in rare instances	Fig. 19
Detailed cross-section of original Dadant hive	Fig. 20
Details of underside of Dadant hive	Fig. 21
Dadant hive open	Fig. 22
The division board of the Dadant hive	Fig. 23
The story and a half Modified Dadant hive	Fig. 24
Frames of the Modified Dadant hive are the same length as those of Langstroth hive	Fig. 25
The Modified Dadant hive has a 40 per cent larger brood comb area than the 10-frame Langstroth	Fig. 26
The Modified Dadant hive is equipped with metal cover and reversible bottom	Fig. 27
The body is dovetailed and has 11 Hoffman frames	Fig. 28
The regular Langstroth body may be used as a super	Fig. 29
Home apiary, where we kept bees over 55 years	Fig. 30
A Dadant outyard. The Poland apiary	Fig. 31
Another Dadant outyard, the Holland apiary	Fig. 32
The LeMaire apiary of the Dadant outyard system	Fig. 33
The Milliken outyard of the Dadant system	Fig. 34
The Koch apiary of the Dadant system	Fig. 35
Another Dadant outapiary on the edge of the bluff	Fig. 36

ILLUSTRATIONS

Queen cells hang downward	Fig. 37
A small colony is confined to such space as it can cover, by use of the division board	Fig. 38
Supers set back for ventilation in hot weather	Fig. 39
The bee-escape board lends itself to modern honey production	Fig. 40
Using mud to close gap between stories by hive tool	Fig. 41
Method of lifting supers to put on escape board	Fig. 42
First honey-extractor of Hruschka	Fig. 43
The first extractor made in the United States	Fig. 44
Uncapping at a single stroke of the knife	Fig. 45
Robber cloth and pan protect supers from robbers during extracting	Fig. 46
The original capping can is still in use in the Dadant apiaries	Fig. 47
Supers returned to the hives after the last extracting of the year	Fig. 48
Super combs which have been in use over fifty years, and are fully as good as at first	Fig. 49
The Mississippi bottoms viewed from the Koch apiary on the bluff	Fig. 50
Chart of the Dadant apiaries in 1880	Fig. 51
Map of the Dadant apiaries in 1919	Fig. 52
Moving bees on auto trucks	Fig. 53
Closing brood-chambers with screens for hauling in hot weather	Fig. 54
Hives packed for winter out-of-doors	Fig. 55
A three-colony winter-case	Fig. 56
Gas torch for singeing hives that have contained American foulbrood	Fig. 57
Low, moist, rich bottom land that is good for honey production	Fig. 58

The Dadant System of Beekeeping

CHAPTER I

Early Experiments.—Natural History

In writing a book, it is customary to begin with a preface, which very few people read. We want the student to read

MOSES QUINBY
Originator of the frame which the Dadants adopted.

this as well as the rest of the book, in order to secure an idea of the "why" of the system that the writer develops in the ensuing pages.

L. L. LANGSTROTH
The inventor of the movable-frame hive

The senior Dadant, Charles Dadant, emigrated to America from France in 1863. He had kept bees as a pastime in Europe. He was very fond of bees, ever since his youth and had already experimented on different styles of bee hives. In the early issues of the American Bee Journal, especially in Volume III, in 1868, he told his beginnings in apiculture.

In 1864, located upon a small brush farm, two miles north of Hamilton, Illinois, he secured two colonies of common black bees in box hives. From these two colonies came the establishment which is now supporting a number of families and has made

C. P. Dadant, sons, and son-in-law

From left to right—L. G. Saugier, M. G. Dadant, C. P. Dadant, H. C. Dadant and L. C. Dadant

them pecuniarily independent. Mr. Dadant often said that those bees always paid, beforehand, in product, for all expenses put in their management and in numerous improvements.

After experimenting upon divers Old World methods of beekeeping, he read in a magazine, of the success of Moses Quinby, of New York State, bought his book "The Mysteries of Beekeeping" and later "The Hive & Honey Bee" of Langstroth, which he was to revise, 20 years later, at the request of the author. In a very short time he became convinced that the Langstroth system was ahead of anything yet devised; that the movable-frame hive principle was the key of successful beekeeping; because of the perfect control which it gives the beekeeper over the bees and the combs.

The difference between the Langstroth and the Quinby methods resided only in the dimensions of the hives and of the frames; Quinby having adopted the Langstroth invention of movable-frames, but adapting it to frames and hives of a different size. This invention consists in hives containing frames of comb with a bee-space between the ends, tops and bottoms of the frames and the body of the hive, opening at the top.

The writer, son of Charles Dadant, was initiated in the main secrets of the bee hive at the age of 14 but did not become an active apiarist till the age of 18.

He was very timid with bees, being afraid of their stings. It was not until a very prosperous season for honey production when, his father being laid up with hay fever, he felt compelled to look after the bees. Finding the hives full of honey and the bees working eagerly in empty supers and filling them with beautiful white combs, his enthusiasm got the better of his fear of stings and he became a bee lover henceforth.

The senior Dadant was fond of experiments. So he tried not only the Langstroth and Quinby styles of hives, but a number of other styles, and in 1868 had a few hives in square frames 12"x13", which he liked well enough to recommend them in a little book, "Petit Cours D'Apiculture" published, in the French language in 1874, for the benefit of his native country. After trying them for 30 odd years, side by side with largely increasing numbers of both Quinby and Langstroth hives, we finally adopted the Quinby size of frames, adding to the number of frames from the original eight, recommended by Quinby, to 9 and 10. Why did we find the Quinby frame preferable? Because it is a little over two inches deeper than the standard Langstroth frame, contains more surface of comb, and supplies more honey over the cluster for winter.

Although we do not propose, in this book, to describe at length the natural history of the honeybee which is given in more or less detail in both of our published works, "The Hive & Honey Bee" and "First Lessons in Beekeeping," it is necessary to explain some of the characteristics of the queens, workers and drones in order to indicate the requirements that bring about the greatest success.

The Queen

The queen, the mother-bee, is fertilized for life, at the age of about 6 to 10 days, in normal circumstances. She is then fitted for a life's production of bees. Her greatest laying comes at the opening of spring, when it is necessary to rear, for the honey harvest, a large number of worker-bees. Early writers assured their readers that a good queen could lay from 200 to 500 eggs per day, and they perhaps wondered whether the reader would believe this assertion. But when the invention of movable-frame hives enabled the beekeeper to study the innermost secrets of the bee-hive, it was found that queens of good quality

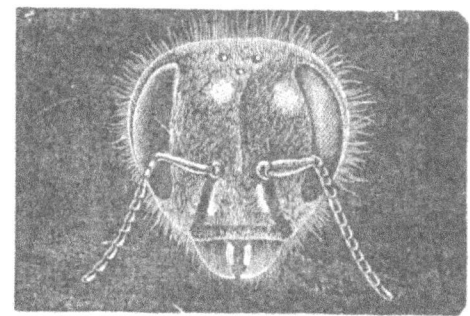

Fig. 4. Head of the Queen (magnified)

(and we should have no others) could lay more than 3,000 eggs per day, for weeks and months together. This was asserted first by Langstroth and Quinby. Mr. Langstroth stated that he had seen a queen lay, in an observing hive, at the rate of six eggs per minute. We witnessed a similar performance ourselves. It is not necessary that a queen should lay eggs at that speed in order to prove very prolific, since a ten hour day of egg-laying would produce 3600 eggs.

Doolittle, one of the bright lights of beekeeping, from 1870 to 1918, asserted that he had had queens that laid as many as 5,000 eggs in 24 hours, for weeks in succession. There is a way by which any one, who owns bees in movable-frame hives, may ascertain how many eggs are laid by a prolific queen, without being compelled to watch her performances. It takes 21 days to carry the newly laid egg, intended for a worker-bee, through the different stages of metamorphosis, to the perfect insect with wings which cuts itself out of the sealed cell. So if we count the number of cells containing brood and eggs, during the height of the breeding season, if the hive be large enough and the queen sufficiently prolific, we will ascertain that many queens can and do lay 3,500 and even more eggs per day, for a number of weeks. To count the number of cells it is only necessary to measure the number of square inches of brood surface, remembering that each square inch represents between 27 and 28 workers.

This heavy brood laying lasts only during the spring and early summer months, of course. In the fall the laying is reduced and in the winter it ceases.

It is important that we should enable the queen to lay to the utmost of her capacity for the time when her bees, or the bees hatching from her eggs, will be able to harvest a crop. Like a good general, we must marshal our forces for the battle neither too early, nor too late. With bees, it is more important than with men, because bees have but a very limited time of usefulness. In order to illustrate this, it is necessary to say a few words about the

Worker-Bee

The worker-bee is an undeveloped female. Had the young female larva, when hatching, been fed with milky pap during the whole time of her existence as larva, and had she been placed in a spacious queen-cell, she would have been a queen. But only one queen is needed in a hive. So the female larvae, and the drone larvae as well, are hatched in small cells and fed with the milky pap, or royal jelly, during only the first three days of life; after that time, the food is coarser and composed of pollen, or bee-bread as it is often called, and honey. The result is an entirely different being from what it would have been as a fully developed insect:

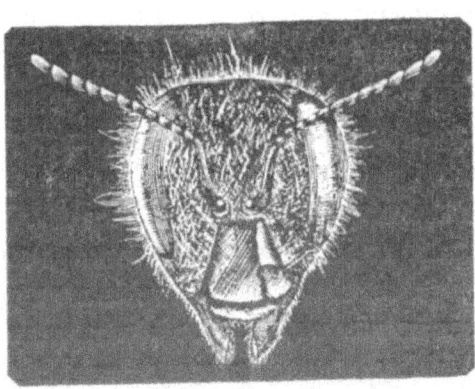

Fig. 5. Head of the worker bee (magnified)

1. The worker has no egg sacks or ovaries; at least such ovaries as she may have are unable to produce eggs in ordinary circumstances and never more than a few eggs at best.

2. Her head is shaped differently, her jaws more powerful, her antennae rise up in the air at the least disturbance, she has salivary glands which do not exist or are very imperfect in the queen and the drone, she has wax-producing organs which exist in neither of the others; on her hind legs are brushes and baskets to gather and carry pollen, which are entirely absent in queen or drones.

3. She is of a determined disposition, flying at an intruder without fear, when irritated, while the queen runs and hides from view. Her sting is straight instead of being curved like that of the queen. She goes about in search of honey and pollen, takes care of the brood and does all the domestic service of the home.

4. She has great reverence for the queen-mother, while the queen herself, if any rivals are about, seeks to destroy them.

The worker bees hatch from the egg in 21 days, in ordinary circumstances. They do duty about the inside for a week, before attempting a flight outside. About the seventh day, in the afternoon, they take their first flight, coming back home after having made the acquaintance of the surroundings, and remaining another week indoors. It is therefore only after 2 weeks of insect life, 5 weeks from the laying of the egg, that the worker becomes a field bee and begins to harvest honey. This is important to know, if we wish to get our force of honey harvesters at the right time.

The worker does not live long in the busy season. Previous to the introduction of the Italian bee in apiaries of common bees, it was difficult, if not impossible, to ascertain how long a worker lives. The introduction of Italian bees has helped solve this problem in a positive manner.

If you introduce an Italian queen in a colony of common or black bees, say on the first of May, after having killed its black queen, there will not be a single black bee left in that hive on the first of August, 3 months (90 days) later. When you killed the black queen, there were eggs freshly laid in the cells. Those eggs have required 21 days to hatch. So in 71 to 72 days from the hatching of the last eggs, the black bees have disappeared, showing an average of 36 days for the life of a worker-bee in summer. During the fall and winter their life is longer, for they do not wear themselves out then with hard work, as they do in summer.

We must compute the time when the marshalling of our army of bees will be needed, by the crops of the country in which we live. We must rear our bees to work upon the bloom of either clover, alfalfa, basswood, sage, heather, mesquite, palmetto, gallberry, buckwheat, spanish needle or bidens, etc., as the case may be. The strength of the colony must come for the opening of those crops; for the general who brings his soldiers too late upon the battlefield will lose the battle.

In other words, to use a most forcible expression, from one of our best teachers among the bee educators, Mr. Geo. S. Demuth, we must rear our working force of bees *for the honey crop and not upon the honey-crop*. The secret of success is all there.

However it is very clear that we cannot have a full force of bees, if the capacity of the brood chamber is inadequate to supply breeding space for our most prolific queens. The queens must not be cramped for room to lay, at the time when bees are most needed.

CHAPTER II.

Size of Hives

The Quinby hives originally made by Moses Quinby and first adopted by us contained 8 frames measuring 10½x18 inches inside. The Langstroth hives of standard size contained

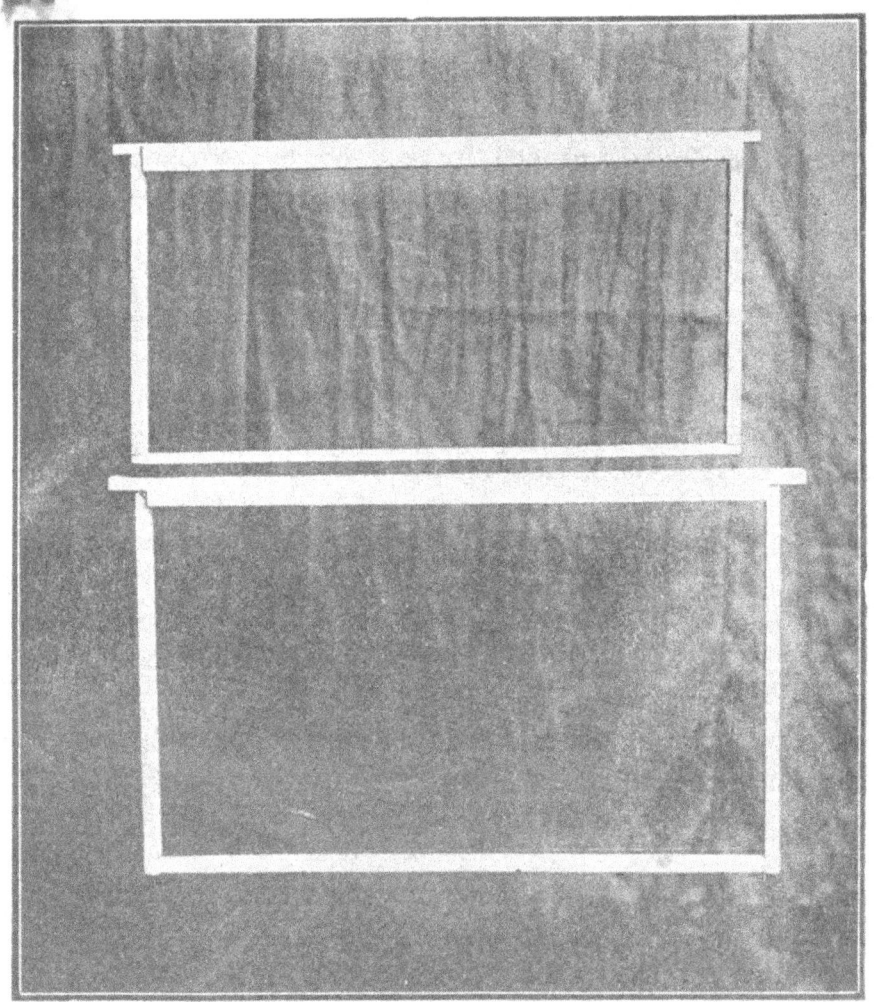

Fig. 6. Frames of Langstroth size and of Quinby-Dadant size

10 frames, measuring 8x16⅞ inches inside. The former hive had therefore a capacity of 92 square inches of comb over the latter. We found both too small, for when we placed supers with

combs over those hives, in the spring, the queens would go up into those supers and lay eggs, after they had filled the lower story. Mr. Dadant senior then tried Quinby hives of different capacities, up to 20 frames of comb, or with a capacity of 3780, square inches of comb. These were too large by all means,

Fig. 7. Twenty frame Dadant hive tried by Charles Dadant

and although we used some 40 of them for several years, we could only produce honey on what was later called "the long idea,' that is to say we had both brood and surplus honey in the same apartment. Although the colony usually occupied one side of the hive with brood, the queen often roamed from one end of the hive to the other and honey sometimes had to be extracted from combs containing brood, a very undesirable arrangement.

To give the reader an idea of the number of hive experiments made by the senior Dadant, we will say that he tried hives with frames 18x18 inches, looking like regular barns.

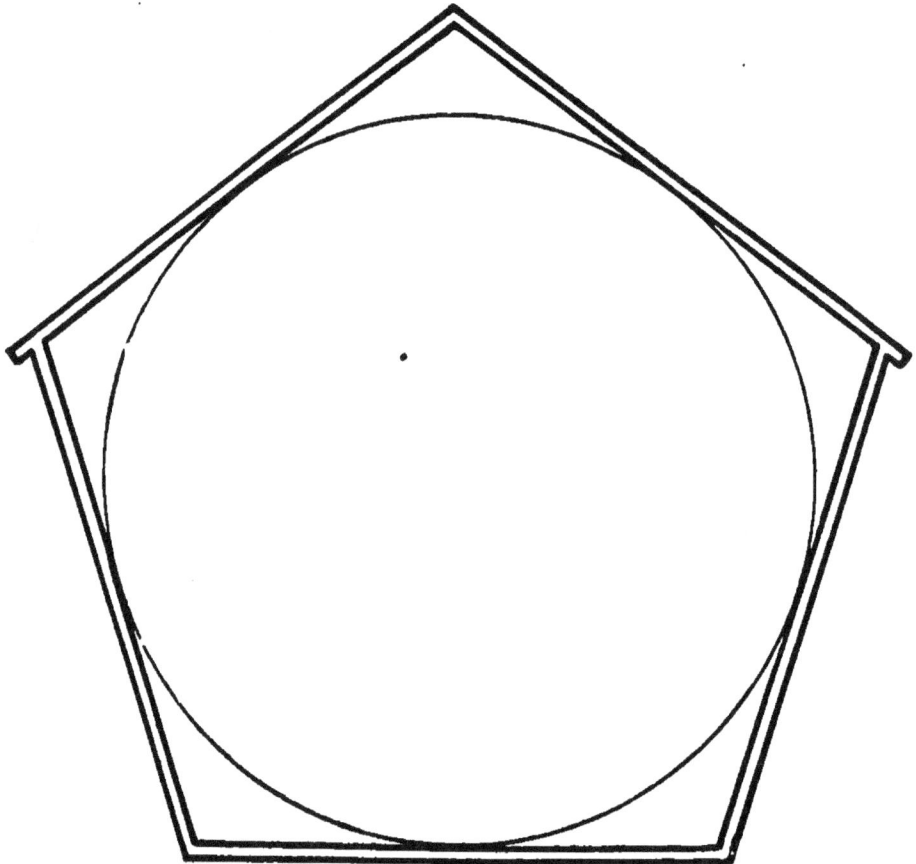

Fig. 8. Coffin-shaped frames tried by the senior Dadant to obtain the nearest to a sphere

These combs proved too large; they would break down readily in hot weather. He also tried hives with frames in the shape of a coffin, because he had noticed that bees rear their brood and cluster themselves in as near a round shape as possible. The coffin shaped frame was the nearest possible to a circle in a hive made of lumber. The bees thrived in them. But the difficulty was in placing supers on those hives. After a few years of trial they were discarded also.

After a number of similar experiments, Dadant senior finally adopted a Quinby hive of 11 frame capacity, reducing it to 10 or to 9 frames or even to a less number for small swarms, with one or two division boards.

The ten-frame Quinby brood chamber which is now called the "Dadant hive," contains 1890 square inches of comb or 540 square inches more than the 10-frame Langstroth. That this brood chamber is sufficient and the 10-frame Langstroth insufficient, for the average prolific queen, in spring, was ascertained positively by us when, about 1876, we handled several hundred of these hives under exactly the same management, side by side with 110 10-frame Langstroth hives, which we had leased for honey production from an old Missouri

Fig. 9. Eight-frame Langstroth hive and Dadant hive side by side

beekeeper by the name of Barlow. During the month of May quite a number of the Langstroth colonies, having been given supers of built combs, began to breed in those supers, while none of the queens in the Quinby-Dadant hives occupied the supers. This was clear evidence that it took more than one story of Langstroth ten frame size to supply a good queen with sufficient breeding room, at the time when we must marshal our forces for the harvest.

Additional evidences of the superiority of a large brood chamber were plentiful when the results were weighed. Aside from the fact that numerous old time beekeepers sang the praise of large hives, even when only logs or boxes, we found that the increased population, from ample breeding room given to the queen, unhampered by divisions or spaces, secured a much increased harvest. Perhaps the most cogent evidence that we can cite is the opinion of a farmer's wife on whose farm we had located an apiary composed of both Dadant large hives and ordinary Langstroth hives in about equal number, managed in the same manner. We were paying these people for the rental of the apiary site, in a share of the crop. The lady, who was a very keen manager, asked us one day why we had brought any small hives to their farm; why we did nor keep those hives at home. She did not think we treated them fairly, for she could very plainly see that the large crops came from the large hives and she gave us to understand that she objected strenuously to our keeping an apiary at their farm in future, unless we kept only large hives there. Although we had seen for ourselves the advantages of large brood-chambers, nothing brought the matter to our notice more forcibly than this avaricious complaint.

However, it was necessary for us to keep some bees in standard Langstroth hives, for very few people have, until lately, been willing to buy bees in such large brood chambers as we use. And yet, for success, especially in running for extracted honey, there is no comparison in results.

It is true that, with Langstroth hives of the standard size, one may use a story-and-a-half or two stories, for brood. But here we found objections. When the queens ascend from one story to another, compelled by necessity, to find room to lay, they hesitate in turning back. We have seen and have often had reports of queens laying in three stories, so that their brood was scattered everywhere in those stories. So the using of a small hive, to be made larger by doubling it, is only using a 'more or less inconvenient large hive. An 8-frame Langstroth hive is too small. An 8-frame Langstroth hive, doubled to 16-frames, is too large. One may do as our venerable and most practical friend, Dr. Miller did, reduce the 16 frames to 8, when putting on supers. This gives the queen ample room to develop her fertility previous to the honey crop, but it compels the beekeeper to make numerous manipulations afterwards. We want a brood chamber neither too large nor too small, capable of accommodating the most prolific queens, in the main apartment, reserving the upper stories for honey. We found it in the hive in question.

Twelve and Thirteen-frame Langstroth Hives

A number of practical beekeepers, who have also discovered the insufficiency of the 8 and 10-frame Langstroth hive, are using hives containing 12 and in some cases 13 frames. These brood chambers are equal in capacity to the 9-frame Dadant brood chambers, but the great number of frames makes them less desirable. Mr. C. F. Davie made the statement, in the American Bee Journal for October 1919, that comparative experiences between 12-frame Langstroth hives and the deeper hives with a less number of frames were all in favor of the latter.

Small Hives

As we have shown in the preceding chapter, we ascertained positively that small hives produce less bees than large ones, because they do not allow full scope to the queen for her breeding. But this is not all. Small hives cause much swarming. Small hives contain a scant amount of honey for winter. Small hives, being less populous, their cluster is in greater danger in winter. Small hives do not enable the apiarist to recognize the best or most prolific queens, when he wishes to select breeders, since they are in a reduced space. Small hives cast more but lighter swarms than large ones. Small hives require queen-excluders over the brood chamber to keep the queen out of the

Fig. 10. Several stories of eight frame hives make too tall a pile

supers. Small hives have a smaller base of support than larger ones, have to be tiered higher for honey capacity, cause a higher climb to the honey-laden bees and are probably more difficult to ventilate than larger ones.

It is true that small hives are cheaper than large ones. But as they often have to be tiered higher and are more prone to cast swarms, it takes a greater number of them to supply an apiary. So the economy is much less than would appear.

Small hives, however, have the advantage of being easier to carry to the cellar and back, easier to transport from one apiary to another, easier for women to handle in making increase.

An advantage in the manufacturing of the shallow hive which has had great weight with manufacturers of hives is that it may be made of narrow widths of lumber, since the standard Langstroth brood chamber may be cut out of 10 inch stock boards, while the Quinby-Dadant brood chamber has to be made of lumber over 12 inches wide, which is now very scarce, or be pieced together with a tongue and groove joint. This has been made less objectionable by placing the tongue and groove near the top of the brood chamber, where it is covered by the strip that supports the telescope cover.

Of late years, in a view of economy, even shallow brood chambers have been made in two pieces. As this use of narrow lumber is likely to continue, there is probably little manufacturing advantage in the shallower hive. We mention it only because of the criticism expressed by some manufacturers.

Many people, bee owners, rather than bee keepers, do not think that their bees are doing well if they do not cast swarms. For such people the small hive is a boon, for the swarms are numerous. Commercial beekeepers view the matter in a different light.

The greatest excellence of the small hive, in the minds of many people, is its low price. But what would we say of a farmer who built a barn insufficient to accommodate more than one half of his live stock, hay and implements? As Mr. Langstroth put it: "Hiving a large swarm of bees in a small hive may be compared to harnessing a powerful team of horses to a baby wagon or wasting a noble water-fall in turning a petty water-wheel."

These are the qualities as well as the limitations of the small brood chamber.

From the experience mentioned above, the writer came to the following conclusions:

A large brood-chamber should be compact, sufficient for all the requirements of the best queens and for supplying all the honey that a populous colony needs for winter and spring, as well as pollen for breeding.

There are several ways to increase the capacity of the Langstroth hive. A 10-frame hive may be increased to very nearly the proper capacity, equal to that of the Dadant, by adding a shallow story super such as we use for extracting. The hive may also be doubled by adding a full story on top.

We have tried both methods. Neither of them suits us. The explanation will be found farther along.

Fig. 11. Shallow brood chambers of the same depth as supers are recommended by some

Shallow brood-chambers, six inches or less in depth, in two, three or more stories are recommended by some. The Heddon hive which was so praised by its inventor about 1886 and created quite a sensation, on account of the undoubted ability of its deviser; the Danzenbaker hive, adopted and recommended also by capable beekeepers and dealers; both of these styles, as well as some others of similar kind and on the same plan as the eke or the nadir of old time beekeeping, have been used and sustained, because the brood-chamber could be "storified" as the British say, to suit the requirements of the best or poorest queens. One of the best commercial beekeepers of Texas, Mr. Louis Scholl, uses such a brood-chamber and succeeds. His

18 DADANT SYSTEM OF BEEKEEPING

example, however, is not followed even among his neighbors, and we believe we know the reason why.

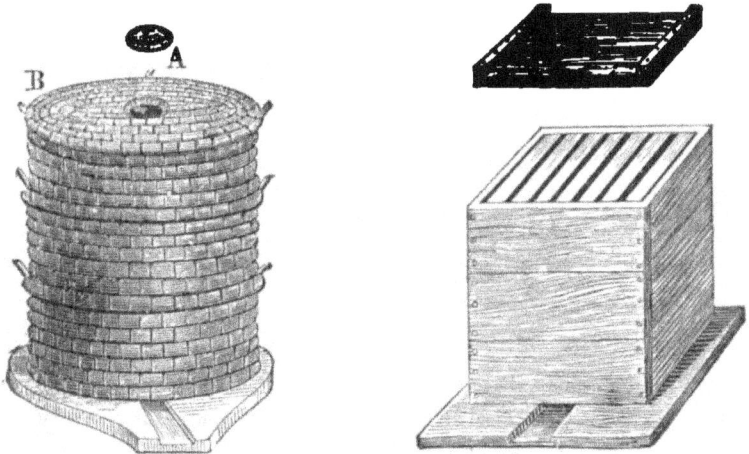

Fig. 12. Ekes or nadirs of by-gone days

In our experiments with bees, we ascertained a fact well-known to the practical beekeeper: the cluster is, as much as possible, in the shape of a ball; the laying of eggs follows a similar plan; the queen begins in the center of the cluster, where

Fig. 13. The DeLayens long idea hive

the most bees are congregated, and lays her eggs in a circle around that center. This is very clearly evidenced when one examines a comb of brood, the older brood always being at the center.

If we stop to think a few moments of the work which is required of the queen, in order that she may lay more than 3,000 eggs per day during weeks in succession, we will readily comprehend that she must not lose much time. Should she lay her eggs without method, here and there, she would be unable to fill the cells with regularity and celerity. When she is about to lay, she thrusts her head into a cell that she believes empty. If that cell is clean and ready for an egg, she inserts her abdomen in it, lays the egg and proceeds to another cell. Without a regularity continued for hours and days together, it would be impossible for her to supply the numerous cells with eggs, so as to leave few empty. She is therefore regular in her actions and goes around the circle with but little loss of time. A very old queen fails in this regularity and should be superseded.

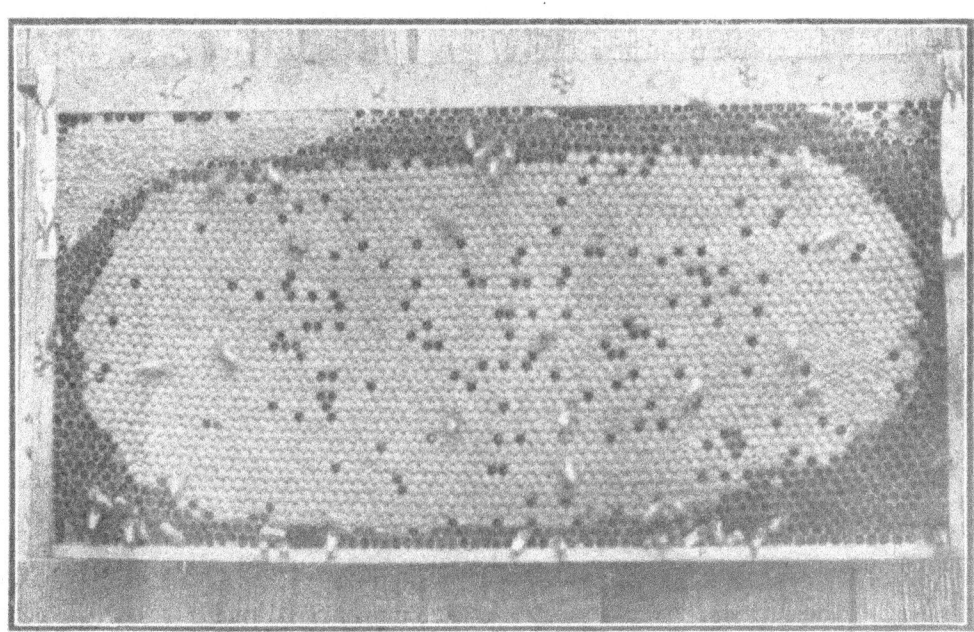

Fig. 14. A regularly laid and well filled comb of worker brood

In a shallow frame, when she reaches the edge of the comb, the queen finds wood instead of cells. This disturbs her and often causes her to retrace her steps and go the opposite way. Even only a cross bar, in a frame, will throw a queen out of direction so thoroughly that she may put brood only on one side of this bar. When she has to go from one story to another she again loses a serious amount of time. When the season is on for her active laying, she is fed so plentifully by the workers that her eggs are produced and protrude from her abodmen, whether she is able to lay them in cells or not. A good queen, in a swarm, will often drop eggs in such number that they may be noticed if the swarm is shaken upon a black cloth. A queen, imprisoned in the hand, during the period of active laying, will often leave eggs between our fingers. So if we would get the best service out of a queen, we must put her in a hive which will give her the greatest facility for finding cells without too much search. This is to be found in a brood-chamber with large combs, where the queen may lay for hours without being turned away from her routine by obstructions of any kind.

When we wonder why a queen lays a greater number of eggs, in a brood-chamber with few combs of large size, than in two or more shallow brood-chambers, superposed over one another, with a bee-space between each of them, the explanation is found in the above statement. The same thing explains why, when a queen has once gone into an upper story to lay, she hesitates to return. She is more likely to go up into a third than to come back into the first. But when she has ample room, on a limited number of combs, to satisfy her propensity to lay, she is much more likely to be contented and there is more egg-laying, a larger increase of population, with less swarming.

This, then, is the explanation of the advantage which we found in results, in those large brood-chambers, as compared with shallow ones.

Safety in Wintering

The advantage of the large brood-chambers is not only in securing larger families at the right time—a large force for the harvest—it is found also in better wintering conditions. The winter cluster of bees occupies a sphere-shaped space, in from four to eight frames, in the center of the brood-chamber. This cluster of bees is perhaps on an average 7 inches in diameter. This means that, in a frame measuring eight inches in depth, the cluster will probably be within an inch of the top. In a frame measuring 10½ inches in depth, the cluster will be 3½ inches from the top. In the deeper frame, there may be 4 or 5 inches of honey, placed by the bees at the top of the combs. In the shallower frame, under similar conditions, there will not be much more than three inches of honey in the same position. The deeper hive is therefore safer for wintering, if our bees are compelled to remain clustered in the same spot for a number of weeks, in very cold weather. As heat ascends, they will be able to eat the honey above them when they would not be able to eat honey at the end of the combs, away from the cluster. This theory, again, is an explanation of the better wintering of bees in the larger and deeper hive, which we found invariable whenever a hard winter made a test of comparative conditions.

Frame Spacing

Another advantage of the large brood-chambers, which we adopted after the example of Quinby, is to be found in the greater spacing of the frames. Quinby spaced his frames—and therefore the combs of the colony—1½ inches from center to center. He followed former apiarists, such as Dzierzon, in this, and thought it the correct distance. Langstroth spaced his combs a fraction over 1⅜ inches apart and thought it correct. The manufacturers of hives of the United States, without investigating the matter very deeply, made the spacing of frames exactly 1⅜. They had good authorities behind them, for Berlepsch, one of the leaders of the middle of the 19th Century, asserted that this spacing was the one followed by bees. Adopting the Quinby hive, we adopted his spacing. When the matter was discussed and we referred to the bees, in natural condition, for their testimonial, we found that they make all sorts of spacing in building their combs, from one to two inches or even more, from center to center. But they aim to have their worker-brood combs a half inch apart, between the combs, which would about represent the 1⅜ spacing, from center to center.

All this may look very unimportant to the beginner. Yet much of the success or failure of beekeeping depends upon just such small matters.

The supporters of the narrow spacing look upon the wide spacing as a detriment and Mr. Julius Hoffman, inventor of the frame bearing his name, wrote:

"If we space the combs from center to center 1½ inches, instead of 1⅜, then we have an empty space of ⅝ inch between two combs of brood, instead of ½, as it ought to be; and it will certainly require more bees to fill and keep warm a ⅝ than a ½ inch space. In a ½ inch space, the breeding bees from the two combs facing each other will join with their backs, and so close up the space between the two brood combs. If this space is widened to ⅝ the bees cannot do this, and more bees will be required to keep up the needed brood-rearing temperature. What a drawback this would be in a cool spring, when our colonies are still weak in numbers, yet breeding most desirable, can readily be understood."

This is a good argument but it does not work well in practice, in large hives. When the breeding season ends, the bees living between combs spaced 1½ inches put more honey in each brood comb, since there is more room and they do not need, then, to keep the cells down to the exact length of the bee chrysalis, as in breeding time. So the comb, or that part of it that is filled with honey, is thickened so as to leave just the necessary room for the bees to pass through. This gives a larger amount of honey than in narrow combs; the bees congregate there in greater number and thus winter better. They are more powerful in spring and the cycle of the year is reached with better success.

The above theory is backed by facts. Bees in large, deep brood-chambers, with the wide spacing, are stronger, winter better, gain strength faster than those in shallow hives, all other conditions being equal. The reader will remember that these facts were established before we tried to explain them by a theory. No theory is strong unless it is backed by the facts of experience. The experience must also be secured on a sufficient scale to make it proof against possible exceptions. The testing of this matter on hundreds of hives, in separate apiaries, treated in similar manner, leaves but little room for doubt of the correctness of the theory. On half a dozen hives, other conditions might change the result.

The spacing of combs, the wider way, has also some influence upon the question of swarming. This will be treated in another chapter.

The Supers

The proper supers to use in beekeeping depend upon the kind of honey that we propose to produce. If we are to run our apiaries for comb-honey, we must have it stored by the bees in a way that will enable us to sell it in the most satisfactory and profitable manner.

When we began beekeeping on a large scale, the little section, made of basswood folded and containing a pound of honey or less, had not yet made its appearance. But sections made of wood, holding from 2 to 4 pounds, had already been produced. Much of the honey was also secured in boxes, the average of them being made of light wood, with one or more sides of glass and of a size to hold 6 pounds, more or less. One or more holes, an inch in diameter at the bottom of such a box, allowed the bees to enter, but did not provide for sufficient ventilation. So the supers used at that date were designed to prevent, rather than encourage the storing of honey.

In producing large crops, we found the bees hostile to small compartments. They would place hundreds of pounds of honey in those large Quinby frames, if available, while they very reluctantly stored a few pounds in small receptacles.

Other beekeepers have found the same thing. Oliver Foster, years ago, wrote:

"When we take into consideration that the object, on the part of the bees, in storing up honey in summer, is to have it accessible for winter consumption, and that, in winter, the bees collect in a round ball, as nearly as possible, in a semi-torpid state, with but little motion, except that gradual moving of bees from the center to the surface and from the surface to the center of this ball, we may imagine how unwelcome it is to them to be obliged to divide their stores between four separate apartments, each of which is 4 inches square and 12 inches long, with no communication between these apartments."

All the experiments made by us led us to the conclusion that we could produce twice as much honey in frames located above the brood combs, without hindrances to the travel back and forth of the workers, the honey to be taken out of these with the honey extractor which had then just been invented, as we had ever produced in small sections or boxes.

But whether to use the double story, or a shallow story for super, on the top of the large brood-chamber, was a question upon which we had to experiment, for very little had been done in this line.

The experiment was made on a large number of hives both with the deep brood-chambers and the Langstroth hives. With the former the matter was settled at once. The brood combs were altogether too large to be used in a super. There was too much danger of heavy combs breaking down, when full of honey, too much trouble in extracting. Besides, a deep super seemed to attract the queen, when a shallow super did not. With the Langstroth brood chambers, the objections to a deep upper story were not so flagrant; yet they appeared to us quite sufficient to condemn it. There was often too much room at one time, so much so that a number of people who double the size of their brood-chamber, previous to the honey crop, often think it necessary to add the second story at the bottom, instead of the top. Then these combs were not so handy as those that we adopted at that time, for extracting combs, with the advice of Mr. Langstroth. Shallow frames, such as have been offered by dealers, are too shallow. Those that we use are of the right size to be uncapped at one stroke of the knife and yet they contain nearly 100 square inches of comb, or over two thirds of the capacity of a standard Langstroth brood-frame. Although we have often heard beekeepers say that they could not tolerate two different sizes of frames in their apiary, we find less objection to these extracting

Fig. 15. Comparison of Langstroth and Dadant extracting supers

combs than any one can find to the use of sections for comb-honey.

The fact that bees prefer large combs in which to store their honey, instead of small boxes, has been also determined by beekeepers in many sections. In Texas, especially, many beekeepers produce what is called "bulk honey or chunk honey," honey in large combs which is cut out of the frames and marketed in tins, with a sufficient amount of extracted honey to fill the interstices between the cut combs. During a visit the writer made in Southern Texas, he was told that beekeepers could expect one third more honey from their bees, in large combs running the full length of the hive, than in small receptacles. This he believed readily, for the assertion was in agreement with his own experience. Aside from the reasons invoked by Oliver Foster, mentioned in a previous paragraph, regarding the habits of the bees, the long comb in the super is more easily reached by the workers, is more easily ventilated, and more homelike in every way. The colony morale—to use an expressive phrase originated by Geo. S. Demuth—the colony morale is enhanced by such supers, the bees work with more entrain and often yield much greater results.

An experiment made by us upon a number of supers, in which both pound sections and full length combs were used, left us not a shadow of doubt as to the bees' preference. Every practical beekeeper knows that bees begin the storing of honey in the super in that part of it which is nearest to the brood, usually the center of the hive. Placing both full length combs and sections, supplied equally with comb foundation, together in supers, but with the sections in the most favorable part, nearest the brood, we saw the bees invariably begin their storing in the full length combs, although remoter from the brood, and they were filled before the sections were fairly begun. Any experimenter can easily test this himself.

Side Storage

In a former chapter, we mentioned very wide hives, with side storage, under what is now termed "the long idea" system. Such hives are used in a number of localities, but the very deep hives are better suited to this method. So the most persistent system of side storage is followed with the De Layens hive, containing as many as 20 or more frames, 12 or 13 inches deep or nearly square. The difficulty is that the queen is at liberty to roam from one comb to another and may have a little brood in most of the frames. Then the honey is difficult to extract. We even tried section boxes and storage boxes in the sides of those hives. The bees prefer to place their honey close to the brood and as much as possible above it. This side storage proved inacceptable, whether in frames or sections.

Queen-Excluders

It is quite the custom for beekeepers to use queen-excluders between the brood apartment and the supers, whether running for extracted honey or for comb-honey. This is because the average beehive is too small in its brood-chamber to accommodate the average prolific queen. She fills it and tries to go elsewhere. She will naturally go up, since the upper story is always warm. The beekeeper is then forced to use some method that will confine the queen within certain limits.

Fig. 16. Queen excluders exclude ventilation and free access to the supers

In these conditions, the queen-excluder comes into good service. It is composed of a frame provided with a sheet of perforated metal, with a bee-space above and below, and placed between the two stories. This perforated metal was, as far as we know, first recommended by a Lorrainer curé, Collin, about the middle of the past century. The perforations are of such size that the worker bees can pass through them, but neither the queen nor the drones can get through, owing to the greater size of their corslet or thorax.

The first queen-excluders made were objectionable, constituting a very serious obstacle to the free passage of the bees and of ventilation. Ventilation is very important, as the bees, in the summer months, are very numerous in the hive. They need pure air as we do, and their great numbers increase the temperature of the hive to the danger point, unless they can force air up through the hive and out again, so as to keep the temperature below the danger point, or about blood heat.

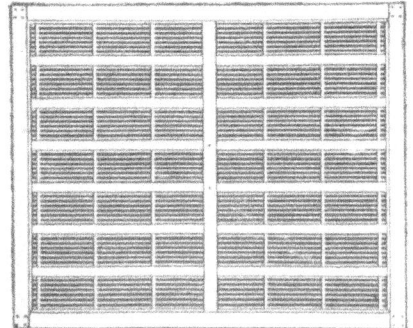

Fig. 17. Modern excluders are better than the old, but are still in the way of the bees

Queen-excluders are now made, with wire spacing, which are much less objectionable. Yet they are still an obstruction to the free passage of air and to the active travel of bees over the combs. Besides, they are expensive, easily put out of service, and are often glued by the bees in such fashion as to make their removal difficult. We do not use them, for these reasons and the following:

With the large, deep hives, it is rare to have the queen ascend into the upper story, unless she is seeking for drone-comb in which to lay and cannot find any in her brood apartment. But if there is no drone-comb in the upper story—and there need not be any if we have used comb-foundation, she will have no inducement to move to the upper story. If she goes there accidentally, she will go back to the lower story readily for the very reason, already mentioned, that a queen likes to lay eggs on large combs where the laying is not interrupted by the obstacles of top and bottom bars and bee-spaces.

It is well, however, to say to the reader that bees do nothing invariably and that there will be queens and occasions, even in large hives, when there will be brood reared in the upper story. But the quantity of this will be so limited that it will never be profitable to use queen-excluders. This again was tested by us on an extensive scale.

In the production of comb-honey, with small hives, on a commercial scale, the excluder is often necessary, for without it the queen may ascend into the comb-honey supers, lay eggs in a few of the cells and thus spoil the looks and the value of the honey for market. With our management, the production of extracted honey and large brood-chambers, we have but one term to describe the queen-excluder. It is a *nuisance*, rarely necessary with our system.

CHAPTER 3
Drones and Drone Production

"The drone" wrote Butler, in 1609, "is a gross stingless bee, that spendeth his time in gluttony and idleness." After 300 years, we cannot find a better description. The drones are the male bees. They are reared in large numbers, in a state of nature, in every hive, because the young queens mate, out of the hive, on the wing, and it is necessary that each queen be enabled to promptly find a mate, in the air. In domesticity, or when colonies of bees are congregated in large numbers in a single spot, the drones produced by one or two colonies are ample for the one hundred young queens that may be produced; since each colony easily rears two or three thousand drones. The drone requires 25 days to change from the fresh laid egg to the perfect insect. He never visits the flowers, goes out only in the warm part of the day, to seek a mate, and comes home with a good appetite, to feed upon the stores gathered by the indefatigable workers.

There is quite a little prejudice in favor of the drones among a certain class of bee owners. They see the strongest colonies produce a large number of drones, then swarm, and they deduce from this that the drone, in some manner, is an influence of benefit to the colony. Some say they keep the brood warm; others think that they encourage the bees to greater activity; in short, they believe that, outside of his functions as male, the drone has a beneficial influence upon the success of the colony. They mistake the result for the cause. Bees rear many drones when they are prosperous, but they are not necessarily prosperous because they rear many drones. Similarly, if we meet a man with a frock coat and a silk hat, we may conclude that he is wealthy, but we must not ascribe the cause of his wealth to the wearing of expensive garments. We take it to be a result instead.

This matter of drone production and of their possible usefulness was thoroughly tested by us. Selecting our best

Fig. 18. Comb built without foundation or with only a small starter. It is practically all drone comb

producing colonies, we supplied their hives with a plentiful amount of drone-combs, so that the queens that we would rear would be sure to mate with good sires. On the other hand, we deprived colonies from which we did not desire males, from the possibilities of producing a large number of drones. This was readily achieved by removing from those colonies all the drone-combs that we could find and replacing them with worker combs. Had we not replaced them with worker combs, the bees would have rebuilt them with drone-combs, as they did, in fact, wherever we neglected to do this. Bees prefer to build drone-combs because the cells are larger and require less material and less time, than worker-combs, so unless the queen desires worker-combs in which to lay, and she usually appears to make her desire known to them, they will usually rebuild the drone-combs that we remove, in the same kind. There are always some drone-cells, here and there, and, in the very best managed hive, the bees will probably rear from 200 to 300 drones. But so small a number is not objectionable. It is the rearing of thousands which is expensive and worse than useless.

It is perhaps necessary to give consideration, in this chapter, to the statement above mentioned, and made by superficial observers, that the drones are useful in keeping the brood warm. This belief was so common that in the old days, French beekeepers often called the drones by the name of "couveuses" (setters); comparing them to the setting hens of the poultry yard. But alas, the drones are reared at great expense by the bees, at the beginning of the season, at a time when the workers could more profitably nurse other worker-bees. When the drones are hatched and begin to suggest to us the possibility of their usefulness in that way, if a cold spell of weather comes, cutting off the honey supply, the bees begin to drive them away, exterminating them without mercy. With the return of warm weather, they again rear a horde of those useless beings, nursing them and coddling them, until the end of the harvest points to them the necessity of again ridding their home of these "idle gluttons."

So it is very clear to us, as it must be to any impartial observer, that the prevention of drone production, in hives from which we do not wish to breed, is in the line of progress. We therefore make it a rule to examine colonies in early spring, and exchange their drone-combs for comb of worker cells. Comb-foundation is not quite so safe; as we have seen, in rare instances it is true, the bees build drone-cells over a worker cell foundation. We have been told that bees will even tear down worker-cells to build drone-cells in their place. This we do not believe, for we have made the experiment of furnishing a natural swarm with a hive full of drone comb and found that the bees were incapable of grasping the possibility of tearing it down to secure worker-cells. They slowly and reluctantly narrowed the mouth of the cells to the dimension of worker-cells and the queen laid worker-eggs in them. There is no more probability of their changing worker into drone cells than the reverse. This experiment was also tried by three leading apiarists of Europe with the same result. They were Messrs. Drory of

Bordeaux, Cowan of London and Bruennich of Switzerland. The results which they obtained confirmed our own.

Although it is difficult to put in figures the economy in honey secured by preventing the bees of a colony from rearing 2,000 or more drones, the attentive student will readily grasp the advantage of the system. Two thousand drones take as much room in the breeding cells as 3,000 workers. Thirty-six drones are raised in a square inch of comb, while the same space will accommodate 55 workers. The amount of food required is similarly larger. But it is after they are hatched as full-fledged insects that the difference in results looms up. The 3,000 workers will be an army of active producers, while the 2,000 gross gluttons, staying home most of the time, get in the way of the workers during the best and most important part of the honey-producing day, from 10 to 4 o'clock.

It is true that we are not always sure of securing 3,000 workers in place of 2,000 drones, for the queen may not lay so actively when she becomes tired; but the economy will show itself so plainly that the beekeeper who tries our method of doing away with drone-comb will never regret it. In fact we believe that the saving in honey, from the prevention of undesired drone production, will be sufficient to pay for a new set of combs every three years.

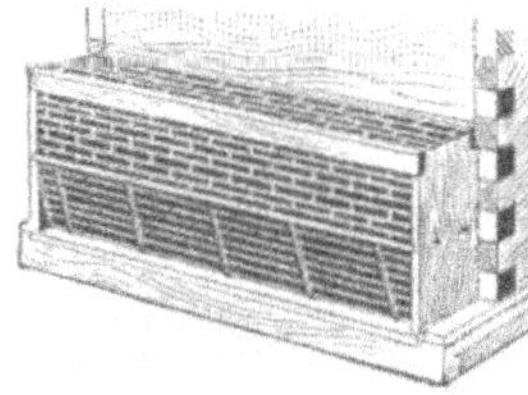

Many apiarists have noticed the objectionable features of over-production of drones. But they have employed, to do away with them, means that were worse than the evil of the presence of the drones. Drone-traps, which every dealer in bee-supplies finds himself compelled to keep for sale, because they

Fig. 19. The drone trap should not be used except in rare instances

are in demand among the ill-informed, are worse than useless. They hinder the bees in their flight, the drones crowd in them and stop the ventilation, the beekeeper is compelled to examine them daily to remove the dead drones, and after all they only do away with the mature drones, when the bees have gone to the expense of rearing them.

Other beekeepers make it a practice to behead the drones when in the cells, before they are emerged. The sealed drone-cells have the peculiar rounded appearance of a revolver bullet in the shell, so they are easily seen. But when we cut off the heads of this brood, we give the bees the unpleasant task of throwing out of the hive all this dead brood, which they pull out of the cells with a great deal of labor. Then, as soon as those cells have been emptied and burnished, there is good chance for the queen to refill them, with the same number of drones, at additional cost to the bees, doubling the work and the expense.

It is therefore very important, and we urge it upon the beekeeper who wishes to succeed, to remove all drone-comb from every colony except those selected by him as breeders, replacing all this drone-comb at once with worker-combs.

CHAPTER 4
The Dadant Hive

In describing the hive which we have been using for years, we will first give it in all its details. We will then show the details which may be dispensed with, in the use of as simple a brood-chamber as it is possible to have while carrying on an extensive honey production.

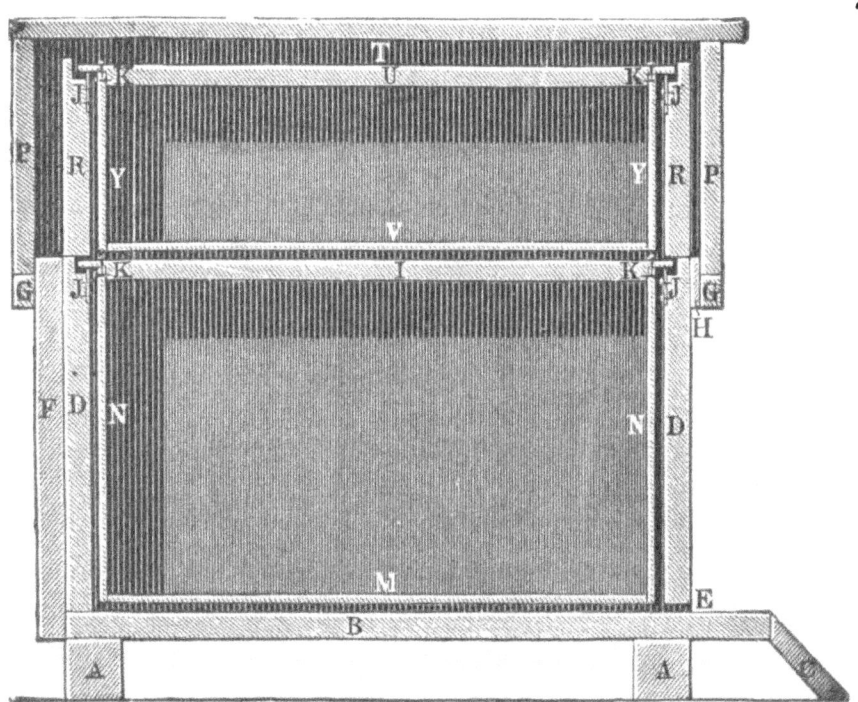

Fig. 20. Detailed cross-section of the original Dadant hive

AA, cross-pieces to support the bottom, 18x2x2. B, bottom, 25x17½x⅞. C, apron, 10x17½x⅞. DD, front and rear of the hive, 16½x12¼x⅞. E, entrance, 8x⅜. F, double board nailed at the rear, 18¼x13x⅞. GG, square slats to support the cover. H, lath, ½x1¾, to widen the top edge of the front board. I, top bar of frame, 20¼x1⅛ wide x⅞ thick. JJJJ, rabbets ½ wide x ⅝ high, dug in front and rear boards, and furnished with sheets of iron ¾ inches wide, or metal spacers projecting ¼ of an inch, on which the frame-shoulders are supported. If the grooves are not provided with these, their size should be ½x⅜.

KKKK, shows how the uprights NN of the frames are nailed to

the top bar. M, bottom bar of the frame, 17-⅞x½x⅞. NN, sides of the frame, 11¼x⁵⁄₁₆x⅞. PP, front and rear of the cap, 18½x9x⅞. RR, front and rear of the surplus-box, 16½x6¾x⅞. T, empty space on top of the surplus-box, 1¼ for the cloth and mat. U, top bar of the surplus-frame, same as top-bar I. V, bottom bar of the surplus frame, same as M. YY, sides of the surplus frames, 6x¼x⅞.

The space between M and B is about ½ inch; between DN, ND, VI, RY, YR, should be ¼ to ⅜ of an inch. Hives of every size can be constructed on this diagram, with the only caution to preserve the spaces of the width indicated. Both top bars are grooved on the under side for foundation and wedge.

The hive body is made with a projection or rabbet, cut into the two side boards, at the bottom, to fit down on each side of

Fig. 21. Details of underside of Dadant hive

the bottom board. The rear is double. The inner rear board fits down upon the bottom while the outer board drops down behind it. This effectually encases the bottom on 3 sides. The purpose of this is to prevent cold air, moisture, moths, robbers, &c., more effectively than with the ordinary bottom.

A wire guide, at the center, near the bottom board, keeps the frames apart. It is more needed in a hive of this depth than with the depth of the ordinary Langstroth hive, but it could be dispensed with, especially with self-spacing frames.

The frame guides, at the point where the frame shoulders rest, in the upper rabbets at both ends, were not used by us

Fig. 22. Dadant hive open; *a*, body; *b*, alighting board; *c*, entrance block; *d*, cap; *e*, straw mat; *f*, oil cloth; *g*, frame of foundation

until we found it necessary to employ inexperienced men in our apiary work. Many old practical beekeepers still dispense with the guides. We have never liked the self-spacing frames. We made a large number of them, in the early days, similar to the Hoffman frames, but their projections were finally all whittled off by us, to secure loose hanging frames. The only advantage we can see to the Hoffman frame is the possibility of handling several of them in one handful. But we never open hives unless we wish to manipulate them and we then prefer to handle each frame separately.

More or less bees are crushed in the joints of self-spacing frames, and where a great deal of propolis is stored by the bees they sometimes stick together unsatisfactorily.

We made the bottom boards with the grain of the wood running lengthwise of the hive. We have also made them with the grain crosswise. We prefer the latter way.

The blocks, upon which the hive bottom is nailed, are useful in keeping the hive up from the earth. When we use cement stands, the blocks are not needed, as pieces $7/8$ of an inch are sufficient. The latter make the hive less cumbrous and lighter. When colonies are taken to the low lands for fall flowers, as sometimes practiced by us, the lighter bottom boards are more convenient. In such cases, some sorts of blocks are used, sometimes only stovewood blocks, to keep the hive bottom up from the ground.

The double back of the hive is for the purpose of protecting the colony better against cold. We aim to face the hives south as nearly as possible, since the north side is always the coldest. The double rear wall on the north, and the division board on the west, we believe to be very efficient in protecting the colony against cold.

The telescoping cover, reaching down $3/4$ of an inch all around the body, and resting on a rim, was adapted from Langstroth's original ideas, as shown in figures, 9, 13, &c., of his original work. The greatest advantage of this telescoping cap is the ability to cover the upper edge of the brood chamber, in an efficient manner. After a few seasons' use of hives, the upper edges of the brood chamber and the lower edges of the supers become rounded and worn so that it is often difficult to prevent robber-bees from securing a passage through the gaping joints. The telescope cap entirely prevents this, is also efficient in covering feeders, placed over the combs of the brood nest, and makes an excellent chaff compartment for winter-packing material over the brood-nest.

A ¼ inch slat is used at the upper edge of the hive front, to widen the projecting edge next to the rabbet. This may seem a very unnecessary contrivance. Yet it is very useful, as it helps us to fit cloths, straw mats, supers, &c., more readily without leaving a smelling place for inquisitive neighbor bees. Those who have kept bees for 25 years or more in the same hives know how annoying the lack of adjustment sometimes is, owing to wear and tear, and to the use of the hive tool in prying apart the stories.

The division board is sometimes called a "dummy." Yet there is a great difference between the two. A dummy is only a board shaped like one of the frames, and of the same size. It fills the place of a frame, but does not conserve heat

Fig. 23. The division board of the Dadant hive

any better than the frame of empty comb. Our division board is closed at both ends, with a rounding piece of oil cloth, tacked on it. It effectually prevents the circulation of air on the ends. Of course, if we were to make it of the exact length of the inside of the hive, we would have great trouble in moving it when necessary, because the bees would glue it fast. But with the soft cloth on both ends, it is moved without jar and without trouble. We remove it temporarily when we need room to handle the frames, in searching for a queen, for brood, &c. It is also moved up to the number of combs actually occupied by bees and honey, when a small colony inhabits the hive. The reader will bear in mind that the hive is made for 10 frames and a division board, while an ordinary Langstroth 8-frame hive, when full, has less capacity than 6 frames of the Dadant hive. We may therefore have occasion to winter a small colony on 6 frames, filling the space behind our division board with forest leaves, or other warm material, and our colony will be more compact and will stand a better chance of wintering safely than the colonies in 8-frame Langstroth hives.

The division board, however, is not made to touch the bottom of the hive, but a bee space is left under it. Our reason for this is that, often, bees have found themselves imprisoned behind a full depth board. Also, in manipulations, it is inadvisable to use a board which may crush bees when put down in place. Our board does not crush any bees and yet serves the purpose of confining the heat of the cluster. Since heat rises, there is but little deperdition of it at the bottom. Not so, with the dummy, which is open on both ends and serves very little purpose. The top slat, or top bar, of this division board, is made of exactly the same size and thickness as the top bar of a frame. It fits in the same rabbet and is not in the way of the supers.

The oilcloth over the combs has proven very superior to honey boards, as the bees cannot glue it fast as they do a board. When we remove a honey board from the top of a brood-chamber, there is a commotion and a jar, for it is always glued. In cool weather, the removal of such a board sets the entire colony in an uproar, as a kick from the operator might do. A cloth is pulled gently from the top of the frame and "peels off" just as far as wanted, without any jar and without exciting the bees.

We have used oil cloths, painted ducking, khaki, gunny cloth, &c. Anything which will confine the bees will do. Some object to these articles because the bees tear them up. It is true that they do so, in the course of time. However, if the cloths are strong and well painted, they last a long time. Their usefulness permits us to put up with their inconveniences. We replace them as often as needed.

The oil cloth is constantly kept upon the hive, except in winter. When putting on supers we place it at the top of the supers.

The straw mat is an implement which very few people use. It is essentially a European implement, cheap and serviceable. European gardeners use mats of all sizes, to protect their cold frames, their hot house windows, their chicken coops or rabbit dens. We make them of the proper size for the top of the frames. They keep out the cold in winter and the heat in summer. In winter, the cloth being removed, the mat is placed directly over the combs, then absorbents over that. It is one of the most economical implements of the bee hive. Chas. F. Muth, an eminent beekeeper of the olden days, whom the writer knew well, used similar mats and considered them very valuable, as we do. But they do not seem attractive to the average beekeeper. Yet they keep away the heat of the sun, in July, as efficiently as the cold of winter, in January.

The telescope cover is made to accommodate at least the depth of one super. It might be made much shallower, but we have found it very handy in covering feeders. When more than one super is on, the telescope cover does not reach down. At that time it matters little whether the joints of the hive and of the different supers are uncovered, for that is the time of the honey harvest and bees do not try robbing then, usually.

The depth of our supers, as will be seen, is greater than the depth of the supers popularly supplied by dealers. This depth was decided upon by us, after due consideration and also consultation with Mr. Langstroth, years ago. At that time, some extracting supers were used of the same depth as the pound sections, i. e. with frames $4\frac{1}{4}$ inches in depth. We considered these as mere playthings, of which too many would be required for the supers of our large and populous hives. We therefore made supers that would supply frames with a 6¼ inch side bar, 6⅝ in the clear. These combs are easily uncapped at a single stroke of the knife. Deeper frames are less convenient and shallower ones insufficient. Later, some manufacturers made supers with 5⅜" frames. We prefer our size. It contains about ⅔ of the quantity of honey contained in a standard Langstroth frame, is much more easy to handle, is uncapped in a single stroke of the knife and runs much less risk of broken combs, when the honey is fresh, the comb new and the weather hot.

For the supers, the spacing of frames is still greater than for the brood combs. Most of our extracting supers contain 9 frames only, spaced about 1¾ inches from center to center. New frames containing comb foundation are first spaced only 1½ inches, as the wider spacing would cause the building of too heavy a comb of honey which would be frail, when new, for handling and extracting. But after the first extracting, such combs are strengthened by the bees and may then be spaced 9 to a super of 16 inches in width. The 1¾ spacing gives us the convenience of thicker combs of honey and less of them to uncap, thus producing more honey with a minimum of combs to handle.

It is well to state here why we make our super only 16 inches wide inside, while the brood chamber is a half inch wider. It is for the purpose of fitting the cover or cap more easily over the supers. The half inch space adds to the convenience.

As to the slogan "Only one size of frame in an apiary," so popular among bee men, we do not think it concerns the supers. It is no more trouble to have full stories for brood and half stories for honey than to have brood frames below and comb-honey sections above. A number of noted apiarists of the special honey-producing sections of the north agree with us fully in this. So will the beekeeper who gives the matter a fair trial.

A Simplified Dadant Hive

First of all, in giving the possibilities of our system with simplified hives, we agree that it would have been best, in adoptng the Quinby depth of hive, to adopt the Langstroth length, which is 1⅛ inches shorter. Not that we believe the Quinby-Dadant hive too long, but because, had we adopted the Langstroth length, many manipulations would have been possible

Fig. 24. The story and a half Modified Dadant hive

with a combination of the two. Thus, although we do not propose to change the length of our hives, after using so many during more than a half century, yet we are quite free to advise the use of the hive called in Europe the "Dadant-Blatt" and in this country the "Modified Dadant."

But in using a style similar to the Jumbo, we do not recommend the narrow spacing of 1⅜ inches from center to center for brood frames as made in the Jumbo. We insist on the style of spacing which we have found very superior, already mentioned, of 1½ inches from center to center, for the brood frames. This was already mentioned in the chapter on "frame spacing." In addition to the advantages mentioned there, we will soon describe another having to do with swarm prevention. The bottom board of a simplified Dadant hive may be made plain, as in the common movable-frame hives sold by dealers.

The hive encasing the bottom board has the great fault, in the minds of many apiarists, to prevent the tiering up of hive bodies, since the projections would keep the upper hive from joining on the lower one and the bees would build comb between them. This is true. But we have found less need of tiering up brood chambers than the average apiarist would imagine. However, we are free to accept this as one of the leading points in the making of a simplified Dadant brood chamber. We have used hundreds ourselves with plain bottom.

Fig. 25. Frames of the Modified Dadant hive are the same length as the Langstroth frames, but 2⅛ inches deeper

Fig. 26. The Modified Dadant hive has a 40 per cent larger brood comb area than the 10 frame Langstroth

The number of frames should be not less than 10, especially in the modified style, which is shorter than the regular Dadant and therefore contains a less amount of surface. The division board should be retained. This means an additional frame space, or an eleven frame width.

The double back is useful only in localities where the bees are wintered out-of-doors, without packing cases. Where heavy packing cases are used or where bees are wintered in the cellar, it is better to have hives single-wall all around. It is an economy and makes the hives lighter for transportation. The encased bottom-board, Dadant fashion, is not indispensable.

Fig. 27 The Modified Dadant hive is equipped with metal cover and reversible bottom

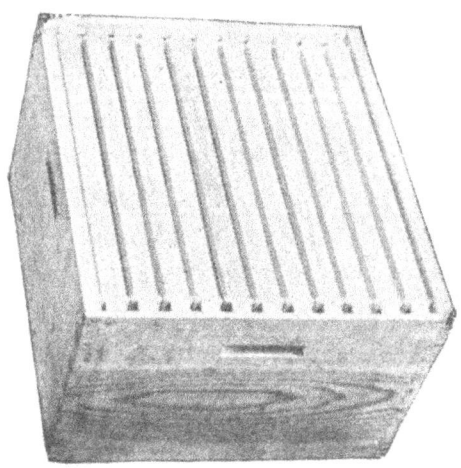

Fig. 28. The body is dovetailed and has eleven Hoffman frames spaced 1½ inches from center to center

We would retain the slat, outside of the upper rabbet, at both ends, as it costs but little and makes the supers so much more readily adjustable. We would also retain the telescope cover, for its usefulness, if its cost were not objectionable.

Hive Making

The writer used to manufacture hives by hand. The fault with such hives is that they are less accurate than those that are

Fig. 29. The regular Langstroth body may be used as a super for the Modified Dadant hive

made by machinery. We remember vividly the annoyance encountered, in the early days, from having employed a carpenter who carelessly cut the hives a quarter inch shorter than the pattern; so the frames of one kind would not fit in the other. It is very important that everything be exactly of the same size and, when we can afford it, we should use factory made hives, or, if we cannot make the hives ourselves, at least employ careful help. We made and used hundreds of hives, cut with a hand saw and later with a foot-power saw. We painted our hives ourselves, also, and did all this work in the dull season. We have always given the preference to white pine for bodies and cypress for bottoms.

As for the frames, it has never paid us to try to make them by hand. Some styles of frames, such as the Hoffman, cannot even be manufactured in a small saw mill, for it takes special machinery to cut them in a profitable manner.

CHAPTER 5
Handling the Bees

We were already keeping bees on a large scale before the invention of a practical bee-smoker. We used a piece of punk or of dry rotten wood, upon which we had to blow in order to produce a sufficient amount of smoke. The writer remembers being often dizzy from blowing his breath upon the rotten wood that he held in his fingers and which gave little enough smoke when the bees were cross. The bellows smokers, invented by Quinby and improved by Bingham, are a greater boon than our younger beekeepers, who have never had to do without them, can realize.

There are men who are either immune to bees or whom the bees do not sting. They handle them without smoke, without veil, and seem to care nothing for angry bees. We were not, and we are not yet, of that kind. The writer was very much afraid of bees in his young days. It was not until an overwhelming honey crop came that he conquered his fear of stings through enthusiasm. So the timid beekeeper should take courage. But we never believed in handling bees without smoke, using it, not plentifully, but judiciously, when opening hives. Many an enemy has been created to the keeping of bees, in suburbs, in villages, along the public highway, by careless handling of the colonies by a beekeeper who is not afraid. He does not get stung, but his neighbors, or the casual passer-by, are the victims. Allow us to relate an incident.

We had a friend, now deceased, in the neighboring city of Keokuk. His home was located on the edge of the bluffs, with no neighbors between it and the Mississippi River flowing below, 200 feet away, an ideal place for the avoidance of bee stings. But he had neighbors above him, up the bluff. He was in the habit of opening his colonies without smoke. The bees never stung him. He told the writer one day that he would have to sell his bees, a dozen colonies or so, because the neighbors complained of stings. He could not understand why they should sting them, when they did not sting him. After this explanation, we walked to the apiary and he opened a hive without using any smoke. Two or three angry bees, instead of stinging him, began to fly about and attacked the neighbor's dog, 40 feet away, up a steep slope. I called his attention to this, advised him to avoid opening the colonies without smoke, and from that day on he had no more trouble with the neighbors. You, inoculated beekeepers, who are careless and think everybody ought to be as sting-proof as you are, take heed of this. Consider other people's safety, if yours is secure, and do not handle your bees at any time without smoke.

We always carry a veil. It may be kept in the pocket, but should be always at hand, to be used in case of need.

We avoid disturbing the bees in cold weather, or in early morning, or late evening. The best time for manipulations we consider to be during the heat of the day, when most of the old bees are busy in the field.

Fig. 30. Home apiary, where we kept bees over 55 years before the publication of this book

CHAPTER 6.

Our Apiaries

Our home apiary, where bees have been kept continuously since 1864, is not in an ideal location for honey. In fact it is only ideal for its sheltered position, in a gentle slope to the southeast, under shading forest trees, which yet do not prevent

Fig. 31. A Dadant outyard. The Poland Apiary

the rays of the morning sun from striking the hives. We consider shade very useful, in the hot summer climate of Illinois. At one time we thought we had too much, as it had become very dense. The bees had had several poor crops and we were inclined to ascribe the inactivity of the colonies to too much shade. But before we had opportunity to cut down any of this shade, (this was in 1903) we harvested one of the largest honey crops that we ever secured. We concluded that it is difficult to place the apiary under too much shade, in this climate. We use roofs, made of coarse lumber, over the colonies, and we believe that it is profitable, both in the economy of wear of the hive lumber and in the shelter it furnishes the bees during the warm season.

In establishing an apiary, we believe it is necessary to place the colonies in rows. But we do not wish great uniformity, great enough that the workers and especially the young queens be unable to easily recognize their homes.

Fig. 32. Another Dadant outyard; the Holland apiary

With 30 or 40 hives, of similar color and shape, ranged in regular rows, without outside guide-marks, there is ample occasion for bees to make a mistake, in their first flight, and enter the wrong hive, on their return home. It is of little consequence with the workers, unless too great a number of them should "drift" from the weak colonies to the stronger. This is likely to occur more

Fig. 33. The LeMaire apiary of the Dadant outyard system

Fig. 34. The Milliken outyard of the Dadant system

or less. But the greater danger is when a young queen, returning from her mating flight, enters the wrong home. She is almost certain to be put to death, if the colony which she enters by error is queenright. Then her own colony becomes hopelessly queenless, unless, as is rarely the case, they still have sealed queen-cells from which they may rear a new queen. At best, the colony suffering from that accident is very much delayed in its breeding.

For this reason, we like to arrange our colonies so that every few feet, there will be some noticeable mark, such as a tree, a bush, or a greater space between hives. It is sufficient to call the beekeeper's attention to this, so that he may go to the trouble of arranging the colonies in such a way as to avoid the "drifting." Painting the hives of different colors is a good way. But dark colors are objectionable, as they are too hot in summer, and light colors soon fade. We know this by long experience.

Outapiaries

We have had outapiaries since 1870. They have been located in all sorts of spots, near the Mississippi River, where the water cut away half of the pasture; in timber land; on the open prairie, among corn fields and wheat fields; on the lowlands of the Mississippi, where only fall blossoms abound; and along the Illinois River, in a similar position. We did not find any ideal spot. There are advantages to each location and disadvantages as well. We hauled bees, in the heat of the summer, on hay racks, a distance of 25 miles, when it was necessary to travel nearly all night to avoid the exposure of the hives to August sunshine. Nowadays, when we transport bees to the lowlands, we use a motor truck and the trip which occupied 8 hours, in 1881, is now made in a trifle over two hours. The younger generation will indeed never realize the hardships that the grandfathers had to face. But our young generations are accepting tasks that we could never have dreamed of and their hardships may prove equal to ours, after all.

It is hardly within the scope of this book to dwell largely

Fig. 35. The Koch outapiary of the Dadant System is located on the edge of the Mississippi bottoms

on outapiaries. We refer the reader to the work "Outapiaries" by M. G. Dadant, son of the writer, whose present experience in that line is quite active. Suffice it to say that we have found, within a distance of four miles, a difference in quality, color, and quantity of honey harvested. This shows that bees, in this locality at least, do not usually travel much over 2 miles in search of honey.

CHAPTER 7.

Apiary Management—Spring

One of the early requirements to which we attend in apiary management is to ascertain that our hives are plumb, from side to side. From front to rear, we want a little slope towards the entrance, so that the bees will have less difficulty in removing rubbish, dead bees, &c. It is also useful in keeping

Fig. 36. Another Dadant outapiary located on the edge of the Mississippi bottoms. The Sack yard

out the rain water and the melting snow. Where we have cement blocks for hive stands, there is less danger of the hives becoming displaced in winter than where they stand on wooden blocks.

The first spring examination is cursory. We make it when the bees are taking one of their first flights. The purpose is to ascertain whether they have sufficient stores to carry them to fruit bloom, and to clean out the dead bees from deceased colonies and close them, so that neither robbers nor moths may enter.

Queenless colonies are not looked for until the second visit, shortly before fruit bloom. The weak, queenless colonies are united with others, lifting their combs bodily, in a cool, evening, to place them behind the division board of a queenright but comparatively weak colony which does not occupy its entire hive and will be benefited by the help. If both colonies have been given a little light food, there is no fight.

We have also found the newspaper plan of uniting bees, given by Dr. Miller, a very good method. It is used quite generally. During a cool night, place one of the colonies, to be united, over the other, separating them with a newspaper and closing all entrances except that of the lower hive. The bees gnaw the paper slowly and generally unite peaceably.

If the queenless colonies are not weak, as some times happens, we give them a queen purchased from a reliable Southern breeder. We used to rear our own supersedure queens and it is a desirable thing to do. But we have been too busy, for years past, to spend any time at this work, which requires special conditions. Southern breeders, if they are active and honest, and there are many such, can rear good queens by the time our bees awaken from winter rest.

As a breeder is not always able to supply queens on short notice, we are in the habit of placing an approximate order, for an approximate time, early enough to get a part at least, of the queens we will need.

Since we have gone into the keeping of over 500 colonies of bees, we have taken less pains to save a colony which comes out of the winter queenless. Yet, taking it all in all, it is not a bad idea to keep this colony going if we wish to avoid having empty spots in the apiary. We have often helped such a colony with a comb of brood, early, then used it as we might use an empty hive, in making a division, late in May. This is hardly profitable, however.

When we wish to examine the hives for the removal of drone-combs, and the replacing of them with worker-combs, which we consider of great importance, in all but the hives which we select as drone producers, and when we have transfers to make, as happens nearly every year, of crooked combs caused by accidents, or of box hives purchased from old-fashioned beekeepers, which we aim to secure in order to do away with box-hive beekeeping, this work is done during early fruit bloom. At that time, the colonies have the least bees and the least honey and are more easily handled without danger of robbing. Methods of transfer of the combs of box hives are given at length in "The Hive & Honey Bee."

In this locality, a long interval of honey dearth exists between fruit bloom and the first yield of the real honey crop, white clover. This is not the case in many other locations, in which some bloom or other fills the gap. During that interval, it is necessary to look after the bees, for they often decrease their laying in a serious manner. It may be necessary to feed them and it is profitable to do so. But this is irregular, so our system of action depends very much upon circumstances. Rarely, the fruit bloom has yielded sufficiently to induce them to continue breeding until the first clover blossoms appear. Our action, at this time, depends upon those conditions. But we find it indispensable, at any cost, to keep the bees breeding. As Mr. Geo. S. Demuth, of the Bureau of Entomology, puts it so clearly, "we must raise our bees for the honey crop and not upon the honey crop." The following of the above axiom, for years, has probably been the most positive reason of our success in beekeeping.

The Honey Crop

We have had a number of seasons when the white clover, whose bloom forms the principal crop of our locality, was entirely killed by the drought of the previous season. In such years, we are pleased if our bees make enough to support themselves till the fall harvest. It is during those years that we aim to make increase, for we have bees in plenty and little for them to do.

Increase

We do not believe in natural swarming. So we do everything that will tend to prevent it. Although "bees never do things invariably," as very truly said by Dr. C. C. Miller, we have very little swarming, little enough to render it unimportant and the watching for swarms a negligible matter. This is of importance to us, since we are constantly busy at a variety of labor, for we are not only beekeepers, but also dealers in bee goods and makers of comb foundation, with a large force of men under our direction. It is also very important not to be required to keep a watchman at each outapiary. It is true that there are seasons when the bees get ahead of us and when it might pay to keep a watchman at each apiary. But these seasons are rare. In 50 years past, we have remembrance of only two, when the bees got so far ahead of us that we had swarms, in great numbers, that were not cared for. A little farther along we will give our method of preventing swarming. We will first give our method of making increase.

We stated, in the chapter on spring management, that we do not now rear our own queens. But as we have reared queens for increase, at one time, we will indicate the method which we followed when neither the Doolittle method, nor the still more modern Barbeau method, were in use. The apiarist who wishes to rear queens, in the most modern way, should refer to special works on queen-rearing.

Queens for Increase

In early spring we remove as much as possible, all drone-comb from the colonies, except from two or three which we desire as drone breeders. These are among our best honey producers of the previous year. We want them pure Italians, because both past experience and breeding theories indicate that hybrids, however good they may prove as honey producers, do not transmit their qualities so invariably as a pure race.

We also prefer pure Italians to other foreign races, because with the Italian we can readily recognize the least amount of foreign blood, while with recognized good races of so-called gray bees, such as the Carniolan or the Caucasian, it is very difficult, if not impossible, to detect a small mixture of the common black bee. The Italians, to our mind, have so thoroughly proved their worth, that we seek no further than pure bees of that race. Imported Italians have been our best bees.

With the method of removing drone combs from undesirable colonies, we secure a very large percentage of good matings, for the colonies from which drone-combs have been taken rarely rear more than 200 to 300 drones, while we secure several thousands in our selected colonies by placing drone combs in the center of the brood nest.

We also select two of our best queens for queen producers. When the time comes, at the opening of what we call the honey

Fig. 37. Queen-cells hang downwards and are built preferable in open spaces on new comb

harvest, we make a good colony queenless and exchange its brood for a less number of combs of brood from one of these best queens. By giving them a less number than they had of their own, we make sure that the queen-cells reared will be well cared for.

To secure a large number of fine queen-cells, we might follow the Alley plan, or the Dolittle plan, but these belong to commercial queen-rearing and we will not describe them. They are described in the Hive & Honey Bee and also in Pellet's Practical Queen Rearing. A very good method is to supply our breeding queens, 3 or 4 days ahead, with new combs or comb foundation cut with rounding edges at intervals, for the easy production of queen-cells by the bees. When these combs are full of eggs and young larvae less than 3 days old, they are just right for our queen-rearing.

Queen-cells hang downward fron the combs. For that reason, the bees, for greater ease, build them at the lower edge of the combs, or in open spots among the brood. If we supply the queenless hive with young brood, less than 3 days old, and eggs, in combs that are fresh and only partly built, there are numerous opportunities for the building of queen-cells in the empty spaces. So a much greater number of queen-cells are built upon such combs. It is nothing rare to have as many as 50 or 60 queen-cells on one or two frames of such comb.

The queenless colony to which these combs are given, immediately builds queen-cells, especially if it is fed with very thin sugar syrup, in case the crop is not yielding. Upon the ninth day, after the insertion of the combs, the cells should be counted and as many divisions of other colonies may be made as the number of cells that may be cut apart, save one which is to remain in the queenless colony. We often divide the queenless colony itself into 3 or 4 parts, closing up the parts that are to be placed on a new spot, in new hives, often carrying them to a cellar or to some cool place, till the next day. We take good care to leave more young bees in the portions removed to new hives and new spots than in the hive on the old stand. We confine them to the space actually occupied, with division

Fig. 38. A small colony is confined to such space as it can cover, by the use of a division board

boards. Other divisions are made in different ways, according to circumstances. We may take only 2 or 3 frames from a full colony, with a large number of young bees, treating them in the same way as those above mentioned. Or we may divide 2 colonies to make one swarm, taking the brood from one and leaving the bees with the queen on the old stand, and taking the bees from the other by placing the division thus made on the stand of the second colony and removing it to a new spot. The colonies that are made from only 2 or 3 frames of brood and bees, will need to be helped later with more brood and perhaps more bees, while those that are built from 2 other hives are at once as powerful as the latter. The 2 or 3-frame colonies, narrowed to such a space as they can cover, with a division board, may be called "nuclei," while the others are at once full colonies. The method of procedure depends entirely upon whether we wish to make much or little increase from the colonies at hand. It also depends upon the question of whether we expect to secure honey from our bees during that crop. If we do, then we divide the lesser colonies into small nuclei, keeping all our best colonies for honey production. We ascertained that middling strong colonies, which may not be ready for the harvest in time, are much more economical for the making of increase, than powerful ones, as many of their bees will be reared "*on the crop*" instead of "*for the crop.*"

The reader remembers that we make these divisions on the ninth day after making our colony queenless. This is because the young queens, ordinarily, begin to hatch towards the end of the 10th day and we must insert a queen-cell in each of those artificial swarms before the end of the tenth day. If we inserted the queen-cells at once, on the ninth day, or waited till the tenth day to make our divisions, many of those cells would be destroyed by swarms that would not have yet ascertained their queenlessness. But on the tenth day, or about 24 hours after making the divisions, our small swarms are brought from the confinement where they were placed and a queen-cell inserted in each, in the middle of the brood combs, in the warmest spot. The young queen hatches promptly, sometimes the same evening, usually within 3 or 4 days, and in another week will be fertilized and laying.

When we make divisions, if we moved the queenless part to a vacant spot and released the bees at once, many would go back to the mother hive or, if too young, might join some queenright colonies; for a new home without a queen, has little attraction for them. Dr. Miller and many others advise leaving the swarm thus made in the apiary, closing its entrance with a bunch of grass.

As we have had several colonies smothered when following this method, we prefer to remove the hives to the cellar overnight. This experience in the smothering of colonies is probably due to the heat of the climate in this locality. By morning, when we remove them to a new spot and insert a queen-cell we have no trouble and the bees seem sufficiently reconciled to the conditions to remain, with the exception of a few of the old field workers who return to the old stand. If the swarm has been made with an excess of young bees, there is no trouble.

Of course, the ideal method is to make one increase from 2 colonies, as all three are strong at once and the damage suffered by the colonies that have furnished this increase is shown only in the lack of honey yield.

It is in either one of these manners that we have increased our colonies, at all times, for we have never liked natural swarming. It is now time for us to state in what manner we prevent natural swarms.

CHAPTER 8

Swarm Prevention and Supering

There are many methods in vogue for the prevention of swarming, but they are nearly all by manipulations which require a great deal of time, at the busiest season. The method which we sustain as best and which we here describe requires no active manipulations during the honey-gathering period, outside of the usually necessary ones, and is what might properly be called a "let alone" method.

As early as 1870, we found ourselves with a sufficient number of colonies to make swarming undesirable. Besides the objectionable increase in numbers, swarming caused an increase of labor when we were busiest. The method which we then adopted has been in constant use by us since, with additional improvements. We do not claim that swarming can be prevented altogether, neither do we claim that it is as easy to avoid it in the production of comb honey as in that of extracted honey. But the success of our management during numerous honey crops is ample evidence that the principles enunciated below are in the right direction. The season of 1916 gave us more positive evidence of its success, as compared to other methods. Out of about 525 colonies, spring count, we gathered less than 30 swarms, but harvested over 200 pounds of honey per colony, while a neighbor of ours, less than two miles from our home apiary, gathered 12 swarms from five colonies, owing to his neglect of proper attendance to their needs. The requirements are as follows:

1. An ample brood-chamber for the needs of a prolific queen. If the queen finds herself confined to a scanty lower story by excluders or otherwise, she will make it known to the bees or they will instinctively notice it themselves and prepare queen-cells. The very large hives, large brood-chambers, and easily accessible supers, that we use, are favorable to a non-swarming disposition.

But even with an 8-frame hive, the prolific queens may be accommodated. Doctor C. C. Miller uses a second brood-chamber for prolific queens and removes this at the opening of the crop, leaving in the lower brood-chamber the best brood-combs. In some way, the queen should be accommodated during the heavy breeding season, and especially at the opening of the crop.

As an outcome of the first proposition, there must be ample room for stores. Some beginners are astonished to see old practitioners, like Dr. Miller, giving their bees as many as three supers at one time, on a strong colony. But if the queen is very prolific, and has been breeding plentifully as nature dictates, her colony may be able to work in each of two or three supers as strongly as they would work in one.

2. The use of comb foundation in full sheets in the supers when working for comb honey, or of fully built combs in extracting supers, has also a great deal of influence upon the prevention of swarming. True, full combs are much more efficient in this, but comb foundation aids greatly. There are days when the crop is so heavy that all the available cells are at once filled with nectar. If the bees have to build combs and thus find themselves crowded for room to deposit their loads, swarming may ensue. But with full sheets of foundation in every section, the work of comb building is much simplified and the necessity of producing sufficient wax reduced. Of course, it is understood that the supers have been supplied to the bees before they found themselves crowded for space, for if the swarming impulse is once gained, it is next to impossible to overcome it by any manipulations whatever.

3. It will be entirely useless to expect the bees to remain contented and fill the supers, if the ventilation of the hive is inadequate to the requirements of the enlarged population. All observers have noticed the great tax imposed upon them by the simultaneous increase of heat and discomfort brought about by a summer temperature and a daily addition of some 2,000 or 3,000 or more hatching bees to the population of powerful colonies. Yet many beekeepers do not think of enlarging the means of ventilation. Thousands of colonies are compelled to leave a large part of their population idle, hanging on the outside of the hive for days and sometimes for weeks, because they

are unable to sufficiently ventilate the inside of the brood-chamber and supers. We must remember that every corner, every story of a hive is in absolute need of a current of fresh air which is supplied at great pains by establishing a line of fanning bees, incessantly forcing pure air in and foul air out. Yet in many colonies there may be but a shallow entrance, partly closed by

Fig. 39. Supers set back for ventilation in hot weather

clustering bees, and perhaps on the inside above the brood-combs there may be some partitions, queen-excluders, separators, honey-boards, etc., all in the way of ventilation. We raise our hives from the bottom, in front, from one to two inches, when there is a likelihood of the bees being unable to ventilate otherwise. We have even set the supers back a half inch or so, during the hottest days, to secure a current of air through the brood-chamber in very hot weather. But this must not be continued too long, for it might interfere with the storing of honey in the forward part of the supers if the weather changed. The bottom ventilation, however, must be ample, ample enough in fact to allow all the bees to work, so that none will remain clustering on the outside during the continuation of the honey crop.

4. As help to ventilation and comfort by decreasing the heat, a good roof is needed when the hives are exposed to the sun. We use coarse roofs on our hives, as stated before, even when they are located in the shade of trees. Our roofs are made very cheaply of large discarded dry goods boxes and are flat. They are cleated with 2x2 inch scantling on the rear underside and a 1x2 inch strip under the front end. This secures a slope of an inch, which may be turned the other way to shed water in rear. The roofs are much wider than the hive and shelter the top from the effects of the weather.

5. The queen must be young. Some beekeepers believe in requeening every season after the honey crop. We do not believe in so radical a measure. In fact, we would not feel capable of killing a first-class queen after only one season of use. But we do believe in keeping only prolific queens and if the queen has proven under grade she should be replaced. Old queens that are losing their fertility are a frequent source of swarming. The workers prepare to supersede them, by rearing queen-cells, just as soon as they notice their reduced laying. The old queen in a pique leaves with a swarm. So we must replace our old queens every fall or late summer.

6. A large number of drones is an incentive to swarming. Some of the old-time beekeepers thought the drones were beneficial because the colonies having many drones swarm readily.

They considered swarming a desirable thing. So it was, when dividing or artificial increase was unknown. They also thought as already stated, that the drones were useful in keeping the brood warm. So they would be if they did not have to be kept warm themselves when they are reared and also if the bees did not kill them, as they are sure to do, in bad weather.

There is not any doubt that the excess of drones in the hive promotes swarming. Those big, noisy fellows remain in the way, all day long, except for a flight during the warmest hours, being then still more in the way of the active workers. Although, as Dr. Bruennich says, there is a certain fondness of the workers for the drones, during the crop, which changes to hate afterwards when they see them helping themselves from their hard earned stores, yet their numbers make for discomfort and a crowded condition.

In a state of nature, according to the best authorities, bees build from one-seventh to one-tenth of their combs of drone size, in the brood-chamber. If only one-twentieth of the combs of a normal colony were filled with drone-brood, this would supply nearly 2,500 drones per colony. We should permit only two or three of our very best breeders to rear so large a number of drones, for 5,000 to 10,000 drones are enough for any apiary.

Some beekeepers see no way to destroy drones but to use a drone-trap. That is to say, during the busiest, warmest season, when their bees need the greatest amount of ventilation, they place in front of the entrance a cage made to catch drones and queens, the very worst encumbrance that may be devised, for the sake of catching the drones as they emerge, having to remove them every evening or suffer the odor and encumbrance of dead drones in front of each hive.

It is probably impossible to rear no drones at all, but if we remove all the drone-comb, early in the season, as nearly as we can, and replace it with worker-comb, there will be drones reared only in imperfect cells here and there or in out-of-the-way corners. Instead of rearing 2,000 or more, we will perhaps rear 200 or less in each colony, a very important difference when we consider the comfort of the colony. Remember that if we leave the bees to their own devices, when we remove the drone-comb in early spring they will be sure to build drone-comb in the same spot. So it is important to replace it with worker-comb.

There are instances, however, of bees building drone-comb on imperfect worker foundation. They are rare and are usually due to some defect of the foundation, which may have been stretched slightly in laminating. At a meeting of the beekeepers of Middlebury, Vt., in the summer of 1916, Mr. Crane mentioned having had about a dozen sheets of foundation thus changed, out of some 2,000 used by him in 1915. These are only accidents. Accidents also are instances of bees building drone-cells on one side of the comb, while worker-cells are on the opposite side. In such a case the regular base is not followed and the cells lap over, showing plainly that they were irregularly built. Mr. Latham exhibited to us two square inches of such comb during the summer of 1916. Such combs should be remelted and replaced by well built combs.

When we replace the drone-comb with worker-comb in all but our best colonies, we do away with undesirable drones for the mating of the queens. We save food which would be wasted on these undesirable beings, since the drone costs at least one-half more to rear than a worker, and has to be fed as long as he lives.

Replace the drone-comb with worker-comb, as much as possible in your hives, early in the season, and you will have much less to fear of the swarming fever.

7. The last condition which we can mention in the successful prevention of swarming is one which we have been using for years, but which we did not think of in that connection until the matter was brought to our attention by Mr. Allan Latham, in 1916. In exhibiting a hive at the Storrs meeting, Mr. Latham made the remark that the 1 3-8 inch spacing of combs, from center to center, in common use, was a promoter of swarming. We have used the Quinby spacing of 1 1-2 inches ever since 1866. The bees work as satisfactorily with the one spacing as with the other. In fact, the original advisors of either mode of spacing had no very positive argument to advance in favor of their

method. But the 1 1-2 inch spacing gives 1-8 of an inch additional between all the combs for the bees to cluster or move about during the breeding season. This multiplied by the height and length of the hive and by the number of frames gives an addition of 162 cubic inches of clustering space or ventilation, as the case may be. Think of the large number of bees which may be accommodated in such a space.

The standard hives of the present day are nearly all of the narrow kind. Nevertheless, the broader spacing is much the better, for the above named reason and also because it gives easier manipulation and more clustering space for the colony in winter. As we have said, we used the wider spacing for years, but did not realize that our success in swarm prevention was in part due to this spacing. It is undoubtedly of great advantage in the prevention of swarming.

Let it not be understood that we lay any claims to the total prevention of swarming. That is a goal perhaps never to be attained. Neither do we lay any claim to breeding a non-swarming strain. But when some of our most practical beekeepers, such as we have met in the East, acknowledge, as one did, having had as many as 18 swarms out, at the same hour, in one apiary, we believe there is need generally of a more thorough understanding of the causes of natural swarming.

The advantages of this method consist in doing away with numerous hive manipulations during the honey crop, such as cutting out queen-cells, taking out brood, shifting colonies, returning swarms to the old hive, etc. All the required work, outside of increasing the opportunities for ventilation and adding supers, has to be done during the dull season. We know that those who have excessive swarming, if they try these conditions, will find themselves greatly relieved by the results. Besides, they may be able to discover additional requirements, for there is always something more to be learned. If we are to judge of future progress by the past, there are endless opportunities for more knowledge, endless chances for progress.

Putting on Supers

We have said enough in previous pages to suggest that we are specialists in extracted honey production. The supers that we use indicate it. We believe that enough more honey may be secured, from combs built previously, to much more than make up for the difference in price between comb-honey and extracted honey.

Every practical beekeeper knows that beeswax is produced in the body of the worker-bee, by special organs acting much in the way of milk production in the cow, or in the production of fat by animals like cattle or hogs. They are also aware that it takes about as much honey, digested by the bees, to produce a pound of comb, as it takes of food to produce a pound of fat in domestic animals. This quantity is not fixed but depends in its proportion upon the circumstances in which the secretion is produced. It is safe, however, to assume from experiments of scientists and from the experience of practical beekeepers, that an average of ten pounds of honey is probably required for each pound of comb. If honey is worth 15 cents per pound, it indicates a cost of $1.50 for every pound of comb.

In addition it is worth while to take into account the loss of time to the bees, when they must remain idle for at least a day to produce this wax, besides the time of building combs. It is true that most of this labor is performed by young bees during the 14 days of their stay in the hive previous to field work. But when a yield comes, it oftens takes old bees as well as young bees to produce the needed wax. So it behooves us to save comb and return it to the bees after having emptied it of its honey, when it has been harvested and ripened by the bees.

Moreover, the production of extracted honey enhances the facility for preventing swarming. Although the method given in the former pages for the prevention of swarming is also successful in comb-honey production, it is much easier to secure an almost total absence of swarms with the production of extracted honey, because we save our super combs from one year to another. We have extracting combs which have been in use for 40 years or more and which are now better than ever, for the bees every season strengthen them by the addition of a little more wax. Neither is there any loss of beeswax in the production of extracted honey. Bees produce comb in large amount when they are compelled to remain filled with honey from one day to another, for want of cells in which to store it. If empty comb is supplied plentifully, a very small amount of wax will be produced. Both practice and theory indicate this.

The supers are placed upon the hives at the opening of the crop of honey. This is perhaps ambiguous to beginners. But it is necessary to have some practical experience to judge of the exact time and even experienced beekeepers may put the supers on too early or too late. We must leave this matter to be decided by the good judgment of the apiarist. We often wait until our best colonies begin to whiten their brood-combs with fresh beeswax at the top. Sometimes, if the crop opens with a rush, this may be too late to prevent swarming preparations. If our colonies are very strong, it does not hurt to put the supers on a few days earlier.

Very strong colonies may be given two supers at once, though we prefer to give only one, waiting to put on the second super until the first one is about half filled.

Supers that are already filled with combs fully built during a previous season are a great attraction to the bees. The producers of comb-honey who follow the method of Dr. Miller and others to place a few already built sections in each comb-honey super, as baits, to attract the bees there, will readily realize what a marked influence supers of comb already built have upon the production of honey. Few people recognize the great cost of comb to the bees. Yet supers are filled more than twice as readily when already full of drawn combs as when given to the bees with only guides or even with comb foundation. Powerful colonies, in a very promising season, may be given as many as 3 supers of the large size that we use, at the very outset, and fill them readily.

Our visits to the bees are timed according to the prospects

of the crop. The weather has a great influence upon results. After several days of rain, during one season, we found a number of colonies of great strength, to whom supers had been given, actually starving. Super combs, partly filled with honey of the previous year, and saved for that purpose, were given them, among the empty ones in the supers. Three days later, the warm sunshine had changed conditions so quickly that we found those same combs of honey heavier than when given. The bees had already gathered a sufficient amount to store nectar in the supers.

When adding second and third supers, we sometimes put them on top of the first, sometimes between it and the brood apartment. If the bees are scattering their honey well, there is no need of lifting a heavy super to put another under it. But if they are rather conservative and crowd their honey into the combs closely, it may be advisable to place the added supers between the stories.

When the crop nears its close, we sometimes find it profitable to equalize supers, so as to find everything full when we remove the honey. For instance, a strong colony may have 3 supers, of which one may be still half empty while the other two are full. Its neighbor may be a colony below average, with one super only, and this only a third full. It is evident then, that the stronger colony will fill its remaining empty combs sooner than the latter. We exchange a few full combs from this strong colony for empty combs of the other, so that, if the crop ends in a week, we will probably find both colonies with full supers.

A super, two thirds full, may be exchanged from a strong colony with that of a weaker one, that has barely begun storing in the only super it has had. Both will probably be found full a few days later.

In these equalizing exchanges, if the crop is on, we do not bother with removing the bees from either super, for bees rarely fight when the new comers are found with full stomachs. In this they are very much like human beings. We smile upon a man who comes to us with gifts; but the beggar, unless he succeeds in interesting us in an apparently true misfortune, gets little sympathy and a cold welcome.

CHAPTER 9

Extracting

We aim to extract the honey as soon as it is ripe, and, if possible, before the crop has quite ended. The reason of this is obvious. It is not so pleasant to remove honey during a dearth, as during a honey flow. But a great deal of the honey must be left upon the hives until the crop is over.

On the afternoon preceding the removal of the honey, we put on the bee-escape boards. This operation is quickly performed with little fatigue, because of the method employed, which does not require the lifting of the supers more than one end at a time, balancing them over a cross piece which supports them until the escape board is slipped on the brood apartment.

Fig. 40. The bee-escape board lends itself to modern honey production

Some persons object to the bee-escape. So did we, before we gave it a thorough trial. Now we have bee-escape boards in sufficient number to supply one for each colony and we find the use of them a great economy of labor. After from 7 to 14 hours, only an occasional bee is left in the super. The only time when bees fail to leave the super is when the queen is with them or when there is brood in the super. They evidently consider it their duty to remain with the queen or the helpless brood and we should not criticize them for it, as it is evidently good judgment on their part.

We have to be careful, when putting on bee-ecapes in times of dearth, not to allow of any passage from the outside for robbers, as they would help themselves to the combs of honey that are deprived of their bees.

Fig. 41. Using mud to close gaps between stories made by the hive tool on supers long in use

Cracks or gaps that may exist between supers, especially when they have seen some 25 years of use, are readily stopped with a little wet clay. The Europeans use cow dung mixed with

Fig. 42. Method of lifting supers to put on escape-board

the clay. It makes a stronger cement. Beginners, with new hives, may laugh at the makeshifts of the practical beekeeper. But the time comes when they find themselves compelled to use similar makeshifts.

It is well, in very hot weather, in our latitude, to avoid leaving those unoccupied supers on the hives during the hot part of the day, as the heat might cause the combs to break down, the bees being no longer there to ventilate and lower the temperature of the super.

If the crop is still on, no precautions need be taken in returning the supers to the colonies. In fact, it is necessary that they be returned as promptly as possible, for the bees are idle meanwhile, if the brood chamber is full. So we make due haste in returning them as fast as emptied of their honey. We saw, during an excellent honey harvest, supers that had been emptied two days before again supplied with fresh honey in every cell. Such harvests are rare, however.

If we have a number of spare supers, we save some trouble by making an exchange, putting on the empty supers as fast as the full ones are removed. Very often, however, in a bountiful harvest, all spare supers have been given to the bees before extracting began. A half dozen, sometimes, are retained from the first colonies harvested, to exchange for full supers as we go along.

If the season is at end, we pile all the supers, as fast as emptied of their honey, in the honey house until evening. A short time before sunset, those supers are returned quickly, everybody helping, to the colonies from which they were taken. This prevents the excitement of their return from lasting long enough to cause robbing, since night prevents further hustling on the part of the excited bees. When morning comes again, everything is quiet.

If the honey is not well ripened, we use honey tanks to store it. But as we have numerous outapiaries and the honey might not be safe in tanks away from the home yard, we use alcohol barrels of good quality and thoroughly dry, to contain the crop. These are emptied at leisure, in vessels of different capacity, as occasion requires.

It is sometimes necessary to haul the supers to the home apiary or central plant, to do the extracting. The writer does not like this method as well as that of local extracting. Yet it undoubtedly has some advantages that cannot be secured in local extracting, because of greater facilities. Each apiarist must decide this for himself. Our principal objection to it is in the necessity of hauling the supers back and forth and the fact that the honey cools and thickens after being removed from the hives and is more difficult to extract. Some beekeepers who use a central plant find it necessary to heat the room overnight, at least, to warm up the honey for extracting.

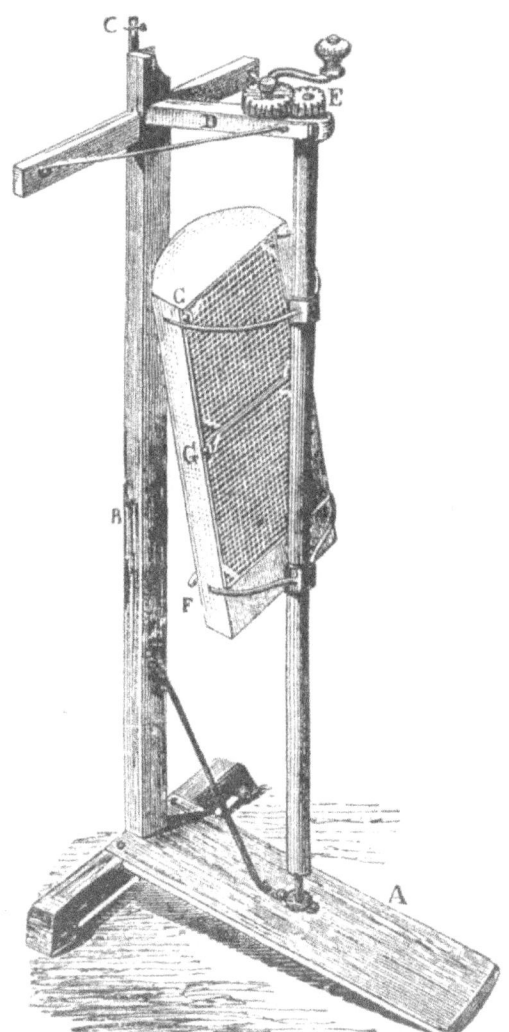

Fig. 43. First honey extractor of Hruschka. The infancy of the development of centrifugal force

Extracting Implements

The senior Dadant was already keeping bees when Hruschka, of Dolo, near Venice, in 1865, invented the centrifugal honey extractor (smelatore). This invention was described in the American Bee Journal, three years later, in April, 1868. It may seem strange to a modern student that so important a discovery should have been 3 years in coming across the ocean, but when we remember that there were only 3 or 4 magazines published on beekeeping at that time; none then in Italy; that the transatlantic cables were just beginning their useful activity, it will be easier to comprehend the slowness of the spread of so useful a discovery.

There were no manufacturers of bee supplies, no dealers in hives or smokers, at that time, in the United States. So the only way to secure an extractor was to have one built at home,

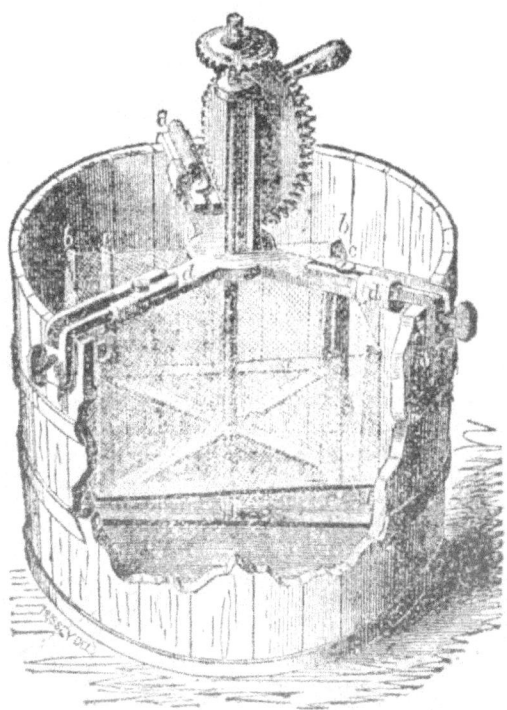

Fig. 44. The first extractor made in the United States was the pattern of our first extractor

after the cut given in the American Bee Journal, Volume 3, page 189. Our local tinner was employed to make the tub, of strong tin instead of wood. The blacksmith made the basket

Fig. 45. Uncapping at a single stroke of the knife owing to the proper depth of super frames

frame, and a churn gear furnished the power. But the fly netting which we employed as screen for the baskets was too flimsy and allowed the combs to bulge in the baskets, so as to break. Add to this the fact that we did not yet have super frames but had to extract from full-sized brood combs, that our combs were not very straight in the frames, since comb foundation was not yet in use, and you will realize how much trouble it was to extract honey in the pioneer days.

But the honey was splendid—what we did secure—so much so that the first dealer to whom we offered it replied very tartly that he did not want any sugar syrup; that whenever he did want some, he could make it himself; that he knew honey when he saw it and had never seen any that light in color.

The coarseness of the extractor was not the only disagree-

Fig. 46. Robber Cloth and pan protect the supers from robbers during extracting

able feature of the newly discovered method. We had no utensils for handling the honey and cappings. A butcher knife, later a thin-bladed knife, served as uncapping knife, and we can still remember the relief brought to the work by the invention of the Bingham knife, whose beveled edge kept the cappings from sticking again to the combs, after having been shaved off. An ordinary dishpan served us for several seasons, as a capping can. This had to be emptied, every few hours, into a sieve with side-boards, draining over a 5-gallon earthen jar.

After 4 or 5 years of this infantile and ridiculous method, or lack of method, we decided to have a large strainer built which would hold the cappings of a whole day's extracting. Manufacturers of honey extractors had then come into being and we ordered from one of them a can of the same size as an extractor can, with another can inside of it. The latter can had a screen instead of a bottom and was ten inches shallower than the outer can, its bottom resting on a pivot in the center of the outer can and its upper edge on the inner edge of it. This was the "capping can," named "Dadant uncapping can" by dealers, which is now often replaced by a long trough with double screened bottom in the honey house. We prefer the capping can to any other implement for holding cappings, because it may be transported like an extractor, because it is light and easily cleaned. Some such implement

Fig. 47. The Original Capping can is still in use in the Dadant Apiaries

should be used. We have several of them, so that we can allow the cappings to drain dry.

We do not like the capping melter sold by dealers, which heats the cappings enough to melt the combs and separate the wax from the honey as fast as harvested. True, it does fast work, but it colors the honey, injures its flavor, and produces too much heat at a time when the weather is already as hot as the apiarist can endure.

We find no trouble in handling the cappings. The honey falling with them into the capping-can drains more readily if the fresh cappings are well stirred about every half hour. This stirring with a sharp, clean, firm slat or stick breaks them up into small bits and greatly aids the draining of the honey. After 24 to 48 hours of draining, they are emptied into a barrel of which one head has been removed. When the season is over, we may have several barrels of these cappings and the entire lot is rendered at one time into beeswax. There is no loss of quality in the capping honey and the wax rendering is done in one operation. Sometimes we wash these cappings preparatory to rendering them. The sweet water thus secured is made into mead or vinegar. Nothing is lost.

The amount of cappings secured from a given amount of extracted honey we find to be about 1 per cent. So when we have a crop of 50,000 pounds of extracted honey, we can figure on approximately 500 pounds of the finest quality of beeswax, after it has been rendered and purified. We used to get a larger amount of cappings. That was when the combs were crooked, or more or less wavy, before we used full sheets of comb-foundation. It was also more difficult to do the uncapping. We use also a less number of combs in the super than in the brood-chamber. At first, when the bees are given the plain sheets of foundation in the super, it is necessary to have the same number of combs as in the brood-chamber. But after they are built, they may be placed farther apart. At each new crop, the bees strengthen them by adding more wax. They also thicken them so that we may use as few as 9 super combs in the super of an 11-frame hive. This gives thicker combs of honey and there is less uncapping. These deep cells are also less likely to attract the queen who requires shallow cells to lay eggs. So the danger of breeding in the super is lessened by the use of a less number of combs in it.

Robbing

We spoke of the possibility of annoyance by robbers, when returning supers of combs to the hives. As a matter of course, these combs are sticky with honey. If the crop has ended, the odor of the honey attracts robbers. That is why we return them to the bees in the evening only. Some people do not return them at all till the following spring. We object to preserving them in this shape, for several reasons.

First. The honey with which they are still slightly smeared is likely to gather moisture, for honey is exceedingly hygrometric, absorbing moisture readily from the air. We at different times had honey to ferment in the cells, during the season following, owing to the presence of a small quantity of this fermented honey in combs that had not been cleaned.

Second. Combs which are sticky with honey are much more attractive to both mice and bees and therefore are more difficult to preserve over winter.

Third. The same trouble is likely to occur when we place these supers on the hive in the spring, as we fear at harvest time. Their strong odor attracts robbers to the hives to which they are given.

Some people allow the bees of the different colonies to clean out the supers, after the crop is over, by exposing them in an open spot of the apiary, until all the honey has been carried out. We have found this to give incentive to robbing and do not like the practice. Besides, if a neighbor has bees, they are likely to come and help themselves also. We believe in being neighborly, but not to that extent.

When robbing has begun on a colony, if it is not worth saving, we break it up and give its combs to other colonies to protect, after having first gotten rid of the robbers.

If the colony is worth saving and we find that it is robbed by only one other colony, we stop the robbing by exchanging

Fig. 48. Supers returned to the hives after the last extracting of the year

the two colonies for each other, placing the robbing colony in place of the robbed one and vice versa. A little flour, sprinkled over the escaping robbers, readily indicates their home.

If the robbing has just begun, it is usually sufficient to throw a bunch of grass loosely upon the entrance of the robbed colony. Its home guards station themselves in that grass as in trenches and pounce upon the robbers who soon give up the attempt.

But it is much easier to forestall robbing than to stop it after it has successfully begun. Therefore we take the greatest care to avoid exposing any honey while extracting.

We use both robber-cloths over the supers while carrying them and tin pans under them, to avoid dropping any honey on the wheelbarrow used, or on the grass, or on the clothes of the apiarist. It is worth while to be careful. One experience with robbing is sufficient to teach us a useful lesson.

While working about the apiary, the apiarist sometimes finds himself followed by robber bees which appear to understand that they may find profit in shadowing him. We have

Fig. 49. Super combs which have been in use for over fifty years, and are fully as good as at first

often stopped our work for an hour or so on this account, or perhaps have gone to work at the other end of the apiary to baffle the robbers.

The worst thing to be feared from robbers is not during the harvesting of the crop. It is when there happens to be disease in the apiary and we find ourselves compelled to open the hives that contain this disease in a period of scarcity of crop. This matter will be considered at the chapter on diseases.

Varying Honey Crops

In our locality, as in many others, the honey yields are divided into two very distinct seasons, the summer crop and the fall crop. The flavor, quality and color of the honeys of these two seasons being very distinct, it is necessary to keep the crops separate. Then also there is sometimes quite a dearth between the two, during which the unsupplied swarms must be looked after, just as we do in spring.

Requeening

It is usually during this short season that we replace the queens which have passed their period of usefulness. Some people change the queens in their hives every season. We believe that a queen is better in her second year than in her first. Exception must be made, however, with the queens that are inferior, by positive test, in their prolificness or in the generation which they produce. Such queens should be replaced as soon as possible.

We allow good queens to become fully two years old before we think of removing them. The removal and replacing with young queens is done with stock purchased from queen-breeders of known integrity, whose breeding stock is known to us and approved as to purity and prolificness.

Rearing one's own queens, from known stock of good honey-producing qualities, is certainly the very best way, even though a beekeeper in the North may find it cheaper in dollars and cents to buy the queens already reared, from a Southern State, where the season lasts longer and bees are not so valuable for honey production.

The advantage of rearing our own queens is that we know exactly what the parentage is, as far as the mother is concerned at least, and if we use a little diplomacy, it is often easy to Italianize our neighbors' bees to secure pure matings. We have repeatedly, in bygone years, Italianized colonies for neighboring bee men, at one dollar per colony, taking all the trouble upon ourselves. We found that it paid us in the long run, as we secured more pure matings; since the neighbors are not so careful as we are in doing away with drone-combs and rear more drones than we do. This is just a suggestion to the careful beekeeper.

Queens will mate with drones from colonies a mile, or two or more away. We believe the drone travels the greater part of the distance.

Some beekeepers rear queens from the very best honey-producing colony in their apiary, regardless of whether the bees of that colony are of pure breed. We do not. Experience has taught us that fixed qualities are very rare in hybrids. Besides, the workers produced by a hybrid queen are usually cross in disposition. Pure Italians have our preference. There are other races equally good, such as the Caucasian. But Caucasian bees are of gray color and a slight mixture with the common race is difficult if not impossible to detect. Italians are very good bees and have the advantage of showing impure blood readily. We prefer them to any other race. We have tried the Cyprian, the Carniolan and the Caucasian. The latter we would take, next to the Italians. The Cyprian are too cross. The Carniolan show too great a propensity for swarming.

Queen Introduction

Some very practical men introduce queens in hives, during the honey crop, by simply removing the old queen, smoking the hive thoroughly for a little while, allowing the new queen to run in at the entrance and closing the entrance for a few minutes. This method has its drawbacks, for it is unsafe in any but a good honey-gathering season. Some people believe in making the colony queenless ahead of time. We don't. We wait till the new queen is there before removing the old one. We believe it a good plan to cage the old queen, for an hour or two previous to killing her, in the cage in which the new queen is to be introduced. It gives the cage the odor of the old queen and appears to make safe introduction more secure. Caging the queen on the brood-comb for a couple of days, we consider the best method of introduction. We have had less losses with this method than with any other.

CHAPTER 10

Nomadic Beekeeping

It sometimes happens that, after the first crop, the weather is so dry that the bloom's prospects are poor for a fall crop.

Fig. 50. The Mississippi bottoms viewed from the Koch apiary on the edge of the bluff

Sometimes also, the first crop is a total failure from drouth. We have had one or two unprofitable years from too much wet. But most of the short crops have been due to drouth.

When the uplands are dry, the bottom lands or low lands near the Mississippi are usually thriving, owing to the early floods which recede in summer. So we practice nomadic beekeeping.

Our first attempt at moving bees to the pastures, in the midst of summer, was in 1881. During that season everything of value to bees on the hills, dried or wilted in June and July.

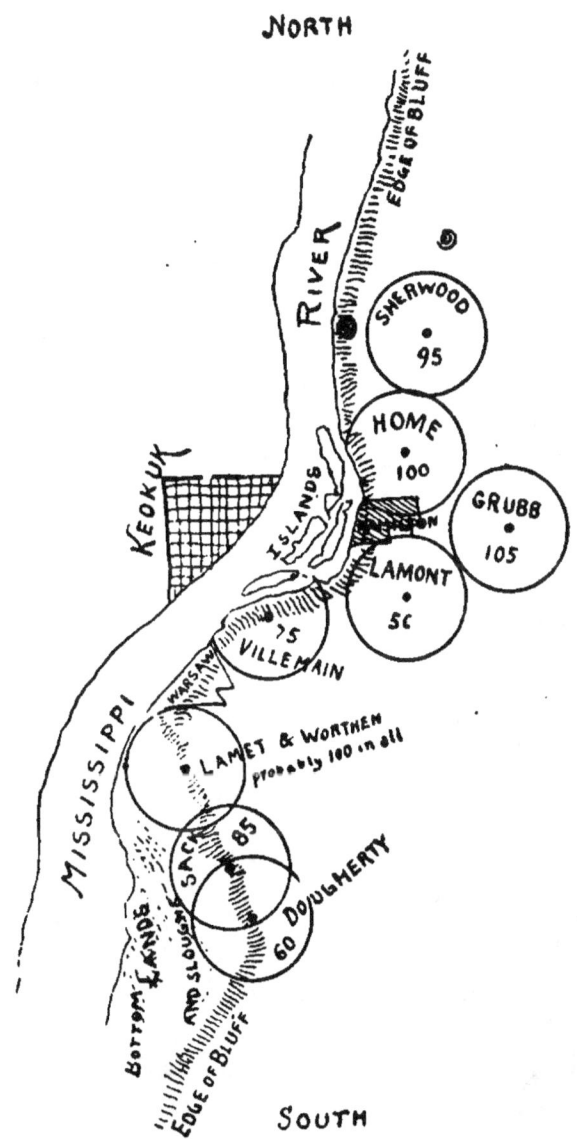

Fig. 51 Chart of the Dadant apiaries in 1880
Each circle has a diameter of 4 miles.

The low lands, having been covered with water until the beginning of July, a very rapid growth of fall plants began as soon as the waters receded. Polygonums or knot weeds, (smartweed) especially of the variety persicaria, also called heartsease; spanish needles or bidens, also called burr-marigolds; boneset and numerous other low land plants, formed great fields of variegated colors, yielding splendid golden honey.

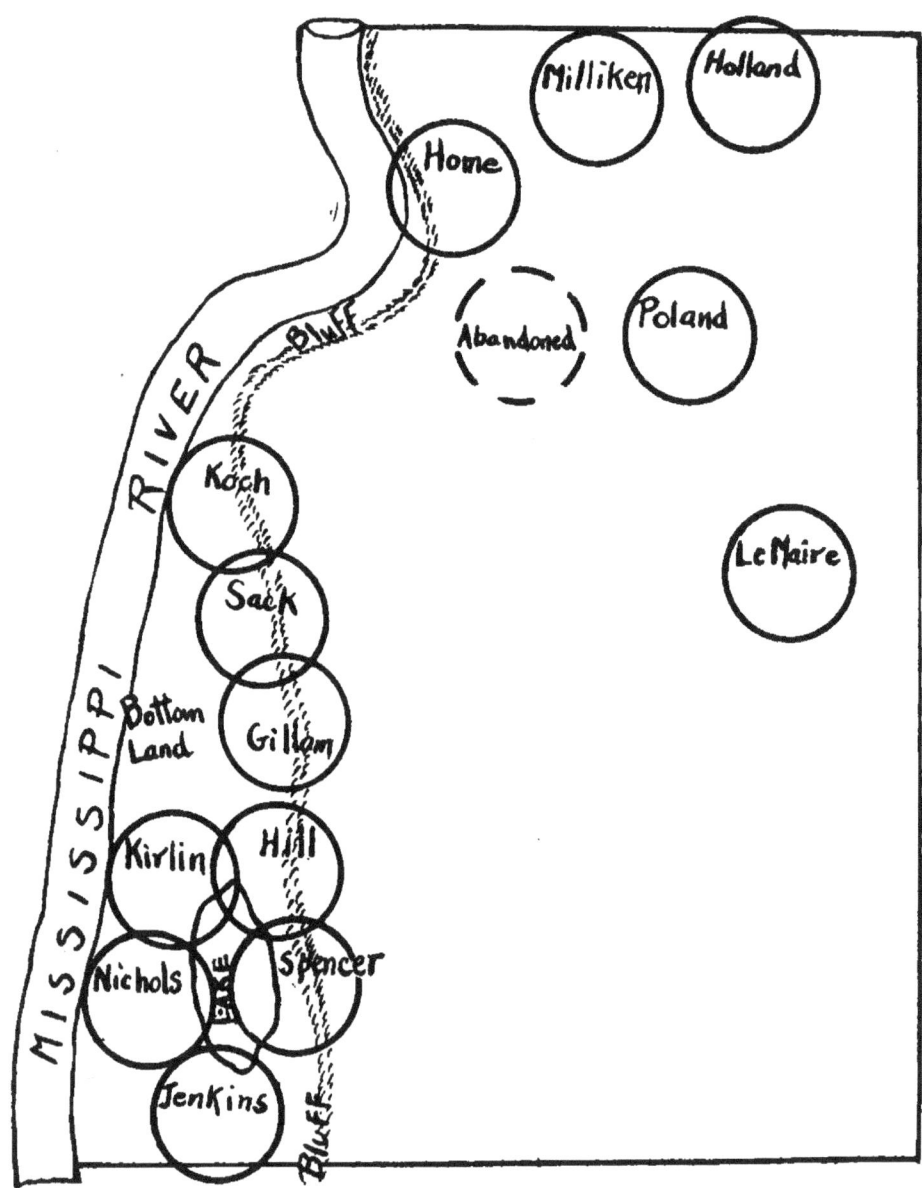

Fig. 52. Map of the Dadant Apiaries in 1919
Each circle has a diameter of 4 miles.

Something over a hundred colonies were moved by us on hay racks, during some of the very hottest weather of that year, a distance of 25 miles.

Luckily the nights were cool. The colonies were so short of stores that some did not even have brood. So, by moving them during the night, we had no difficulty in transporting them. The trips took about 8 hours, including nailing them up and opening them on arrival. The result was very satisfactory and proved profitable.

No total failures happened again for a number of years and it was only in 1918, 37 years later, that we again resorted to moving bees in summer. But at this time, we had auto trucks and we were not compelled to travel at night, though most of the hauling was done early in the morning. In 1918 and 1919

Fig. 53. Moving bees on Auto Trucks

we thus hauled a large number of colonies to the low lands (400 in 1919), with profit.

The advent of the auto truck has certainly made a revolution in nomadic beekeeping, as a man can readily enough haul even rich colonies from one place to another. To confine them, we usually entirely nail up the entrance, as we find that the old bees are apt to worry on the screen, if screen is used to close it up, since that is their usual exit. We remove the cap and the supers, hauling them separately, and each colony is covered with a wooden frame entirely screened over, with cross pieces

Fig. 54. Closing brood-chambers with screens for hauling in hot weather

to protect the screen. Thus 400 colonies were hauled from 25 to 30 miles without any loss. They were brought back to the original apiaries in the fall during frosty weather. At that time no ventilation is necessary. A deep wire cover gives them sufficient air.

In addition to the advantage of bringing bees to rich pasture, we believe there is a gain in what Mr. Demuth calls "morale" in the behavior of the bees, when they are transported. We have often noticed that bees brought to a new spot, whether it be in spring or fall, and having to learn a new country, become more active. It seems to increase their energy, their morale, to find themselves in new fields. We are not the only ones, besides Demuth, to have noticed this. Others have spoken of it also. It sems to act upon them as upon pioneers who go forward to new countries, for we believe that the great energy, the positive activity, of the American people, are due to their having been pioneers in new spots as yet uncultivated. Bees may be induced to more activity in several ways and we believe this is one of them.

A similar argument is brought forward by a beekeeper of the Netherlands, who, after having described the nomadic methods of Hollandish beekeepers, who move their bees to the heather fields every summer, asserts that the moving of the bees increases their activity and the breeding propensity of their queens. We might cite similar opinions from several other noted apiarists.

CHAPTER 11

Fall Management

When the end of the crop is near, we find it advisable to make sure that our bees are putting enough honey in their brood combs for winter.

With comb-honey production, it is often noticed that the bees have crowded so much honey in the brood combs that they have little room for breeding. The colony decreases in numbers, especially if the crop lasts to a late date.

With our management in the production of extracted honey, large brood chambers and very free access to a number of supers, it may happen, if we are not on the lookout, that our bees will have a contrary experience, an insufficient quantity of honey in the brood combs. We believe in a very large amount of stores for winter and spring. The old way was to figure on 25 pounds. We want 40 pounds at least to make sure that there will be no shortage and no stint in the spring breeding. To obtain this, we crowd the bees for room; that is to say, we allow them a little less super room and this forces them to put a sufficiency in the brood combs. Better have 50 pounds than only 25.

It is not advisable, however, to cramp them so that the fall breeding will be neglected. A colony wintering on mainly old bees and a limited number of young bees is not likely to succeed as well as a colony that has bred freely till the late fall. This question requires judgment on the part of the apiarist. Even a very experienced man will make mistakes in this matter. Much also depends upon the conditions at the end of the crop. In many seasons the amount of brood reared, before cool nights come, depends upon the field conditions.

CHAPTER 12

Wintering

If we have a sufficiency of honey in the brood combs, a good strong colony of bees upon those combs, we have the best possible conditions for winter, provided the honey is of good quality.

We had very expensive experience with honeydew, harvested in July and retained in the brood combs for winter. We had similarly great losses with fruit juice, apple juice, grape juice, etc., harvested by the bees in September-October, during a dearth of honey. The season of 1879 was especially discouraging, for, while the bees did no damage to sound fruit and only gathered grape-juice from bird-damaged grapes and sweet cider from damaged apples, many of our neighbors thought that we were getting rich at their expense, supposing that our bees were transforming all that fruit juice into good honey. The truth of the matter was that many a bee was unable to even reach home with the fermented juices that it gathered, in lieu of good honey. Although we tried to take out of the comb all the cider and fermented grape juice that they thus stored in the cells, there was enough left to poison them during the cold days.

Honeydew, fruit juice, cheap syrups, and honey containing many pollen grains, are bad winter food. To keep warm, the bees must consume honey and they must have as pure saccharine matter as possible, as all grades of sweet food containing acids, vegetable matter, starch, etc., leave a large amount of residue in their intestines, which they must discharge out of the hive, on the wing, or suffer from diarrhea or dysentery. This is discharged in the hive, upon the other bees of the cluster, if the weather is too cold for them to void their feces outside. That is why many of the most practical apiarists feed their bees the best quality of sugar syrup to complete their winter stores. It is a good method to follow, even if some desk-educated apiarists affirm that bee diseases are due to the use of sugar in bee food. We are quite willing to grant that, for spring food, nothing is better than honey, because of its containing matter which helps build the body of the young bee. But when the bees are full grown and confined to the hive, nothing is better for them than the purest saccharine matter. So we have always made it a point to remove all honeydew and late-gathered, dubious sweets, harvested during a shortage of flowers in fall.

If the bees are short, we add to their stores by supplying them, in inverted can feeders, in the month of October, with sugar syrup made of 2 parts of high grade cane sugar to 1 part of water. If they do not take it too fast, they will assimilate it, or in other terms invert it, by their saliva, just as they do with pure nectar. If there is danger of the sugar not being sufficiently inverted, it is a good plan to add to the food about one fifth of its weight in good honey, *with the quality of which we are acquainted*. This point is important, for it would not do to use unknown honey that might bring them disease, such as foulbrood.

In our entire experience, we have had 5 or 6 winters when the amount stored by the bees was insufficient to carry them through. Whenever the amount of good sugar syrup given the bees, in addition to their own stores, was sufficient, we regularly found them to winter better than if they had used only natural stores. So we are very much in favor of sugar feeding whenever the colonies are short for winter.

Whatever be the conditions, we aim to have our bees in proper shape for winter long before the opening of cold weather. Just as it was necessary to get them to breed plentifully until the chilly days come, so it is necessary that they should quit breeding when there are no more bees being lost from overwork or accident. We want plenty of young bees for winter, but we do not want them to have an amount of brood to care for, when winter opens. At that time, the quieter they are, the better it is for them and ourselves.

Cellar Wintering

We experimented largely on both cellar and outdoor wintering. For 15 or 18 years, we wintered the bees at our home apiary in the cellar and occasionally also at an outapiary. We discontinued it because, in our latitude, there are too many winters when a number of warm days make it undesirable to have the bees in confinement. We are just a little too far south for regularly successful cellar wintering. If we could foresee the weather, we would put them in the cellar for the cold winters and keep them out-of-doors during the mild ones.

Our House Cellar

In 1875, we built a house in which a hot-air furnace was placed. We separated a space, measuring 12x22 feet, from the main cellar, with a double-wall partition. The parts of this room that had the outer wall exposed to the weather in winter were also lined with boards and shavings on the inside, so as to avoid the sudden changes of temperature. This bee-cellar, when too cold, was readily warmed by opening the partition door that separated it from the main cellar. In case of too much heat, two windows could be opened and in fact remained open most of the winter. But both the air currents and the heat were easily regulated.

In this cellar we wintered bees for some 15 winters, with very good success. The failures were always due to mild weather, when the sun would permit the outdoor bees to fly and our cellared bees were restless. Ice was tried to cool the cellar but it did not seem very practical. We finally abandoned cellar-wintering altogether as unnecessary in our latitude. We have seen many cellars, in locations further north, and we think the hillside cellar, with a well-protected entrance, on a level with the apiary, is the ideal cellar. It must be deep enough in the hill to be immune to temperature changes.

The degree of temperature that we considered the most desirable in the cellar, was from 42° to 45°. Messrs E. F. Phillips and Geo. S. Demuth, of the Bureau of Entomology, section of beekeeping, at Washington, have made extensive experiments and they hold that a temperature of nearly 50° is requisite. It would be very unreasonable to differ with them, after the positive tests and extensive experiments made by them. The temperature, as recorded by us in the cellar, might admit of 50° at the top of the hives. But there is a very safe way to adjust the advice to give the novice. Have a thermometer, try to find the degree at which your bees are quietest, and hold it at that. Everybody agrees that this is sound advice. When the bees are happy, we can just hear a slight rumble, a "bruissement" as the French call it, much resembling the quiet rustle of the leaves among the trees or the whisper of the waves on a distant shore.

Wintering in Clamps

We also tried wintering in clamps, long ago. For two winters, we put our bees in a clamp, just as we would potatoes or cabbage, with the only difference that we had draft holes, for the ingress and egress of ventilation. This method would certainly be successful in cold climates. It failed with us when we had a wet winter, the ground of the silo being soaking wet during the greater part of the winter. The bees suffered from mould and came out in bad shape.

Wintering Out-of-Doors

This was the final method adopted by us. We tried colonies in sheds, sheltered by closing the shed in stormy weather and opening it on warm days. This method was very satisfactory. But sheds for hundreds of colonies are not practical. We tried putting large dry goods boxes over some hives, during the cold. This was also good, provided we uncovered our colonies in warm days.

We tried what has been called "chaff-hives," with 2 inches of chaff or sawdust all around, above and below the hive body. This was good but entailed the making of very expensive and cumbrous outer-cases, which made the hives difficult to transport, especially as our hives are already very large. A chaff-hive of the Dadant pattern weighed 80 pounds, empty. In addition, during one winter we found that the bees which were con-

Fig. 55. Hives packed for winter out-of-doors

fined by the cold needed a flight as often as the weather warmed sufficiently. The colonies in our chaff-hives failed to take flight whenever there happened to be just one warm day, because it took more time for the warmth of the atmosphere to reach the cluster, than with single-wall hives.

We finally adopted the method which gave us the least amount of expense, while securing a good average of wintering, for the winter climate that we have in our locality. The hives are double on the back, as noticed in the description. The brood chamber is reduced to the actual capacity of the colony, by removing all empty frames, if any, and moving the division board to reduce the space as much as possible. Then forest leaves are used both in the body and in the cover, as protection. The oilcloth is first removed, and the straw mat laid directly

Fig. 56. A three-colony winter-case

over the frames, so that the moisture from the bees may escape gradually through it into the leaves of the cap, without causing a draft of air.

The hive is then wrapped with forest leaves, held in place with wire netting of the kind used for poultry yards. A piece of netting of the proper size is laid behind the hive, leaves spread upon it and both ends brought together in front, which is always faced south, southeast or southwest. In this way the hive is efficiently protected from fierce north winds, but the front is open so as to permit the rays of the noon sun to strike it fully. Whenever a warm day comes, usually about once or twice a month, the bees are able to fly, void their excrements and often rearrange their cluster.

Heavy snows are not objectionable, if they do not cut out the ventilation by thawing and forming ice about the openings. We usually see that the snow is piled freely behind the hives, with an open entrance when a thaw comes. Many people object to letting the bees have a flight in snow time. It is true that many get lost on the snow, unless the weather is quite warm. But we have always noticed that the hives whose bees get the freest flight in snow time, turn out as good as the best.

We never disturb them, for any purpose whatever, during cold weather. Even if we wish to pile the snow on the north side of the hives, we go at it very cautiously, so as not to awaken them, as the few bees that would leave the cluster might be chilled and lost.

Many people believe in what is called "sealed covers" over the brood combs. We don't. We tried this to our heart's content. During the winter of 1884-5, one of the hardest that we ever saw, every colony which had absolutely sealed covers over the cluster, was pitifully soaked by the melting of the ice which formed from the bees' respiration. Wherever the bees had covers in which the moisture could escape slowly through the absorbents in the cap or cover, the colony wintered in much better condition.

This test was made by us accidentally. We had on the hives, right over the top of the frames, oil cloths some of which were more or less damaged by the gnawing of the bees. This was neglect on our part. But as it happened, we found the neglect to have been beneficial. For wherever the oil cloth was perfect, the upper escape of moisture was prevented. The moisture settled over and around the bees, congealed there, to thaw out and soak them as soon as the weather moderated. In the hives where the oil cloth was more or less damaged, the moisture escaped among the leaves of the upper story, the bees remained dry and conditions were much more satisfactory. Thus, the poorer the "sealed cover" the better the success.

Let it not be understood that we want the bees to have an open super above their combs. No, we try to have for them what we want for ourselves in a cold winter night, a warm, woolly, moisture-absorbing cover, that will at the same time retain the heat. We have no use, either for ourselves or our bees, for an impermeable cover that will confine moisture near the body.

We do not urge the adoption of the above system of wintering by any one. In wintering, perhaps more than in any other part of the management of the apiary, much depends upon location. Let each man judge of this for himself.

Outer winter cases of different models have been recommended. Some are made to contain 4 colonies; others 3, others still only one. If our opinion is desired, we gladly recommend the 2-colony winter case which does not require the moving of the hives of bees for winter. Two hives may be kept all summer in close proximity to each other; so that they may be incased together without trouble. Whenever we move any hives of bees for winter, we cause more or less of what is called "drifting." Drifting is the loss of direction by some of the bees, because of a disturbance. Whenever some of the colonies are moved, there is drifting. The strongest colonies then gain the greater number of the drifting bees, because they make a greater call, a more discernible noise, which draws the lost bees. So the weak colonies are weakened and the strong colonies strengthened by the "drifting." Very expert beekeepers, with large practice in the use of the packing cases, have acknowledged to us that drifting was one of the main drawbacks of that method.

If the packing of colonies, singly, in thick packing boxes, in a way that they could not at all feel the cold of winter, was not so expensive, it would be the ideal way of wintering bees, even in very cold regions. We have not adopted this method because of the great expense it entails. Our method has been sufficient and although we have lost heavily in a few abnormal winters, we succeed quite regularly. Our losses are not over 4 or 5 per cent, one year with another.

CHAPTER 13

Diseases of Bees

If any one had asked us, 20 years ago, how much trouble might be expected from bee diseases, we should have probably shrugged our shoulders and answered that they were very insignificant and hardly worthy of notice. For 40 odd years after we began beekeeping, the only disease we saw in the apiary was diarrhea, also called dysentery, from which the bees suffer more or less after a protracted winter, especially when their food is not of the best. Spring dwindling, which is a result of this diseased condition, is of importance only in rare seasons and late springs. There is less of it with strong colonies than with small hives.

Foulbrood, in either of its two different forms, was entirely unknown to us. In 1903, the writer had to go as far as Colorado to be able to see some rare samples of it, rare because the beekeepers who had it in their apiaries kept it under control by constant fight.

It was not until the spring of 1908 that we found the disease among our own bees. We had been feeding them some very fine western honey, which we had on hand, so as to avoid buying sugar, thinking that the bees would return that honey with ample interest in a very few days. Whether this was the cause, or whether the American foulbrood which had been noticed in many parts of Illinois had just reached us in some other way, we found it very much scattered among our bees, although none of the colonies had more than a few cells of it.

We did not hesitate in treating the bees. We transferred every colony by the method recommended in text books of the most modern date. We will give the method in a few words.

We go to the first colony, remove it from its stand and put upon that stand a clean empty hive with frames containing only a few starters of foundation. All the bees and the queen are shaken into it and treated just like a natural swarm. The contents of the hive are removed, the brood is burned, the balance of the combs rendered into wax, the honey being heated for at least a half hour to the boiling point of water. When the hive has been emptied, it is singed with the flame of a tinner's gasoline torch and, after supplying it with frames, it is then ready for the transfer of the next colony. This work succeeds best when it is done, as we did it, just at the opening of the honey harvest, and with the greatest care.

After 48 hours, the bees are again shaken onto full sheets of foundation, or on combs from healthy colonies if such may be secured. This method is called the "starvation method." By this is meant that the bees are deprived of their combs and placed where they must build combs. So they consume all the honey in their honey sacks, producing wax. This method was entirely successful on the disease which is called "American foulbrood." All the text books describe it. But we might as well give a short description of it here.

In American foulbrood, caused by bacterium named "Bacillus larvae," which was discovered and named by Dr. G. F. White of Washington, the unhatched bee dies just about the time when it is sealed in its cell by the workers. Its body decays, turns to a brown coffee color, has the odor of joiner's glue, and when a toothpick is inserted into it, it strings out, like so much liquid rubber, to the length of 2 or 3 inches. These 3 symptoms, when together, indicate positively the existence of American foulbrood.

We succeeded fully in doing away with the disease and afterwards harvested the largest crops that we ever secured. But since the disease is in existence in the country, we find it necessary to be constantly on the alert and treat, without delay, any colony in which it is found, even if only a few larvae are diseased.

Neighbors of ours, who made light of the disease when it first appeared, saw their colonies absolutely ruined in a very short time. Two or three men near us quit the bee business entirely, from discouragement. Yet beekeeping was never so profitable as it has been since it requires greater vigilance than formerly.

We also had a fight with European foulbrood in one of our

Fig. 57. Gas torch for singeing hives that have contained foulbrood

apiaries. This mysterious disease, caused, according to Dr. White, by a bacterium which he named "Bacillus pluton," attacks the bee larva when it is small and still coiled in the cell. Some beekeepers say that it has a very disagreeable odor. We have never detected much odor, perhaps because we never allowed it to develop to a dangerous point. We found that the best remedy for it is that given by Alexander years ago and tried with success by Dr. C. C. Miller, removal or caging of the queen for from one to three weeks, during which time the bees clean out the disease, if the colony be strong enough. Weak colonies have to be united. Italian queens have proven safer than black queens in producing a progeny capable of coping with the disease.

Here we must give extra warning to the beginner concerning the possibility of "robbing" when there is foulbrood in the apiary. No combs should be exposed, no honey allowed to be robbed, no bees permitted to enter other hives, for in each of those instances there is danger of transmitting the disease. Thus foulbrood should be treated at the beginning of a honey crop. If we find ourselves compelled to treat it at other times, it must be done when the bees are quiet, when there are no robbers about.

When American foulbrood is treated properly, with the greatest care, there is but little danger of its reappearance except from outside of the apiary. Not so with European foulbrood; for it seems to reappear mysteriously.

We also met with "sacbrood," a disease of the larvae in which they die and dry up, so that they may be shaken out of the cell. This usually disappears of its own accord, when the good season comes. It may be due to faulty queens.

We no longer fear bee diseases, though they are unpleasant to meet. Vigilance is necessary, it is true, but in nothing can we expect to reap a reward without labor. It is quite probable that the time will come again when the foulbroods will be as rare as they were 50 years ago. We read in the text books of the first half of the nineteenth century, that some of the leading scientific beekeepers, such as Dzierzon and Berlepsch, had their apiaries well nigh destroyed by the diseases before they found methods of cure. Earlier still, Schirach, in the Eighteenth Century, discovered that the cure of American foulbrood was to be found in depriving the bees of all honey and compelling them to use that which was contained in their stomachs, because honey is the most active transmitter of what is called "American foulbrood."

There is a very destructive disease of the adult bees called Isle-of-Wight disease, in the British Isles. But traces of a similar disease have done very little harm in this country. However, a disease called paralysis, vertigo, May disease, etc., appears from time to time in different parts of the world. We have seen it, but only on a very limited scale, and it usually disappears with the beginning of the honey crop.

For a more thorough description of all the bee diseases we refer the reader to our larger work "The Hive & Honey Bee."

But let us give one warning: Do not feed to your bees unknown honey, no matter how fine it may be. There is no doubt that honey may be entirely healthy for human beings and yet contain enough of the bacilli of foulbrood to contaminate any colony to which it is fed. Pay no attention to the men who tell you that sugar is not good for bees. If you have to feed the bees and have none of the honey of your own production, better give them syrup made of the very best sugar. There will be no danger of disease

CHAPTER 14

Enemies of Bees

We have found no enemies of bees quite so bad as our own selves, when we are neglectful or too avaricious. We have occasionally removed too much honey, to our sorrow later.

The moths have never done much harm in our apiaries, because we promptly accepted Mr. Langstroth's statement that there is no more danger of a healthy colony of bees being destroyed by the moths than there is of a healthy cow being killed by carrion flies; although a queenless colony will be just as readily riddled by the maggots of the moth as the body of a dead cow will be consumed by the maggots of carrion.

We keep our empty combs from year to year, during the winter in a cold honey house, where the thermometer often goes down to zero F. ($-18°$ C). Neither larvae, eggs nor moths can withstand such low temperatures. In the summer, if the honey house is kept well closed, there is still no danger. But, in the summer, we usually have all our combs in use where the bees care for them better than we could. Hives in which the bees have died at the opening of spring are usually the most fertile breeders of moths, for moth worms often winter in out-of-the-way corners, kept alive by the warmth of the bees. Were it not for such accidental wintering of the live grub, moths would become well nigh exterminated in these northern countries. Such combs should be treated several times during the spring and summer months and it is better to keep them out of the bee house. However, if the treatment is thorough, by burning a quantity of brimstone in an earthen dish, with the supers and combs piled in broken tiers, so as to allow free circulation of the gas produced, the quantity of brimstone that will kill the flies in the room will also kill the moths. The quantity to be used must depend upon the size of the room and its greater or less ability to retain the fumes.

We also use carbon-bisulphide, poured upon a rag and laid at the top of a pile of supers, closing the pile carefully to allow the evaporating gas to penetrate in all the crevices. Care must be taken not to bring a flame near, as this gas ignites and explodes.

Other enemies of bees are too insignificant to find a place in this short treatise.

We will here give a few rules that we consider important, in our climate at least:

We never face colonies northward, as the bees often get chilled near the entrance to their home in unpleasant weather.

We never move colonies of bees, even a few inches, without placing a slanting board in front of the hive, to cause them to notice the change of location.

We never disturb colonies of bees in cold weather.

Fig. 58. Low, moist, rich bottom land that is good for honey production

We never try to winter a colony that has less than 6 combs. Such a colony might be wintered in the cellar, if it has sufficient stores. Smaller colonies are united.

We never close colonies to keep them from flying because of snow or rain or other reasons, except to transport them. We consider that enough bees worry themselves to death to make up for those that would leave and get lost in the cold or in the rain.

We never use drone traps, moth traps, separators, queen excluders, entrance guards, etc. Separators are useful in comb-honey production, but not with our system.

We never cut our queens' wings. We do not disapprove of the practice, but we do not find it necessary in our system.

Conclusion

In our description of the Dadant System of beekeeping we do not wish to be understood as advising the average beekeeper to change from any system that he now uses. There are tens of thousands of beekeepers using 8-frame Langstroth hives, 10-frame hives of the same depth, and even other patterns, with good success. Our own success, for more than a half century, has caused inquiry as to our methods. That inquiry led to the publication of this work. We have no apology to make, but neither do we urge the following of our system nor the adoption of our hive. The only thing which we consider *absolutely indispensable* in modern beekeeping is the use of a movable-frame hive of sufficient capacity for the prolificness of the queens. But there is no doubt in our mind that a maximum crop of extracted honey, can be secured by the Dadant system with less manipulation than with any other system yet made known. More bees may be kept, more honey produced with the same hours of labor. We have given our reasons for using this method. That is enough. Let the reader decide for himself as to his course.

BOOKS ON BEEKEEPING FOR SALE BY AMERICAN BEE JOURNAL

Langstroth on the Honeybee
REVISED BY DADANT

This book, originally written by Rev. L. L. Langstroth, the inventor of the movable frame hive, has been revised and kept up to date by the editor of the American Bee Journal. It is the one book that no beekeeper can afford to be without. It contains careful and accurate accounts of the life and habits of the honeybee and the mysteries of the hive. Full and reliable information concerning the detection and treatment of disease, the sources of nectar and pollen, and care of the apiary throughout the year are included. The best methods of producing and marketing large crops of honey are made clear. This book is nicely bound in attractive cloth cover and contains 575 pages. The price is very low for a book of this size and quality. Price $2.50. French edition, $1.75. Spanish edition, $2.00.

First Lessons in Beekeeping
BY C. P. DADANT

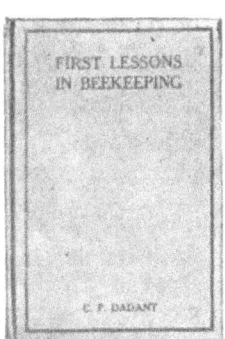

The senior editor of the American Bee Journal, who is the author of this book, has spent nearly all his life in a beekeeping atmosphere. His father, the late Charles Dadant, was an investigator who became well known on both sides of the Atlantic. As a young man, the author of this book was associated with his father in honey production and assisted him in the many experiments which he conducted in his efforts to make beekeeping a practical success.

Contains just the things you want to know, in a style easily understood and with many pictures to explain the text. You may safely recommend First Lessons in Beekeeping to your friends.

167 pages, cloth bound, well illustrated. Price, $1.00.

American Bee Journal, Hamilton, Ill.

BOOKS ON BEEKEEPING FOR SALE BY AMERICAN BEE JOURNAL

Scientific Queen Rearing
BY G. M. DOOLITTLE

An old work that has had a big sale. Gives Doolittle's methods of queen rearing by artificial grafting. We advise "Practical Queen Rearing" as preferable, but the student or commercial breeder who desires to practice cell grafting will find this work interesting.

Price, cloth binding, $1.00. Leatherette, 50c

American Bee Journal
Edited by C. P. Dadant and Frank C. Pellett
Questions answered by DR. C. C. MILLER

Oldest Bee Journal in the English Language. A 36-page Monthly Magazine

Subscription $1.50 per year. Canadian postage 15c; Foreign 25c extra

Every phase of beekeeping is covered in the Journal, every section of the country receives attention. The market page alone is worth several times the subscription price to beepers with honey for sale.

New methods, latest news, illustrated articles on honey plants, free legal service department, questions answered, profusely illustrated. First and best in its field.

If the American Bee Journal is wanted in combinaton with any one of our bee books, add $1.25 to the regular price of the book and both book and Journal will be sent postpaid.

American Bee Journal, Hamilton, Ill.

BOOKS ON BEEKEEPING FOR SALE BY AMERICAN BEE JOURNAL

Practical Queen Rearing
BY FRANK C. PELLETT

In preparation for this book Mr. Pellett visited many of America's foremost beekeepers and queen breeders, both north and south, and has described their methods fully.

The methods of the older queen-breeders and writers, Alley, Doolittle and others, are explained with the variations which are the development of later years.

Simple methods of rearing a few queens in a small apiary, as well as methods used for rearing queens in wholesale quantity, make the book valuable alike to the ordinary beekeeper and commercial queen breeder.

Cloth binding, 105 pages, 40 illustrations. Price $1.

American Honey Plants
BY FRANK C. PELLETT

The first book in the English language on the subject of honey plants

A knowledge of sources of nectar is fundamental to to the success of the beekeeper, as the difference of a mile or two in distance often doubles the returns from the apiary on account of better pasturage.

This book is the result of years of study and visits to important honey producing sections, from New England to California, and from Canada to Florida and Texas.

An authoritative book by an expert beekeeper and reliable naturalist. Profusely illustrated with original photographs.

300 large pages, 155 fine illustrations, cloth bound. Price $2.50.

AMERICAN BEE JOURNAL
HAMILTON, ILL.

BOOKS ON BEEKEEPING FOR SALE BY AMERICAN BEE JOURNAL

A Thousand Answers to Beekeeping Questions
BY DR. C. C. MILLER

For over 25 years Doctor Miller has answered questions for beginners and veterans alike through the columns of the American Bee Journal. More than 10,000 of these questions have been answered in this manner. These have been sifted and more than 1,000 of them included in this new book, edited by Maurice G. Dadant.

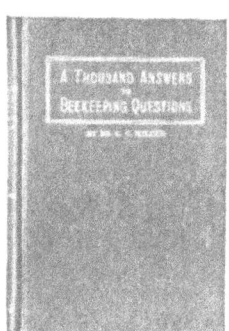

Alphabetically arranged by subject, this book will clear up many problems not touched by other bee books.

The texts all tell a connected story of bee life and the principles of honey production, while this takes up singly the many questions which perplex the beekeeper in every-day practice about his bees.

Should be in every list of bee books.
Attractive cloth cover, 276 pages illustrated. Price $1.25.

OUTAPIARIES
BY M. G. DADANT

A clear and concise explanation of the requirements for proper placing, arranging and managing of outapiaries.
Too many beekeepers expand into outapiary beekeeping without fundamental knowledge of its requirements. The result is that apiaries are often located improperly and have to be moved after errors are discovered by costly experience.
Special chapters are devoted to apiary sites, basis of placing the apiary, systems of management, moving, autos and trucks, honey houses and equipment, and treatment of apiary during different seasons of the year, with special apparatus used by large beekeepers.
This book is especially valuable to the beginning outapiarist, but will contain many items of value to the experienced outyard man.
The book is cloth bound, has 125 pages and 50 illustrations, and is printed on fine paper. Price $1.00.

American Bee Journal, Hamilton, Ill.